http://GISforSurveyors.com

Montana Technical Writing

COVER DESIGN BY Tulasi Zimmer
COVER PHOTOGRAPH BY PATRICK N. SMITH, PLS
ISBN: 978-0-9888737-3-5
PUBLICATION DATE JULY 20, 2013
PUBLISHED BY MONTANA TECHNICAL WRITING

CONTENTS

PREFACE

The purpose of this book is to help the land surveyor understand how Geographic Information Systems (GIS) technologies support land surveying activities. I also discuss the important ways that surveyors support GIS data development, integrity, and spatial accuracy. I present GIS issues that affect surveying, and surveying issues that affect GIS.

This book presents concepts, overviews, and a few specific examples on a variety of topics related to Geographic Information Systems relevant to land surveying, its myriad issues and applications. The intention here is to present ideas to point the reader in the right direction. More detailed training on specific topics is available elsewhere – some of which can be found on the companion website http://GISforSurveyors.com.

Today, GIS is a ubiquitous set of data, tools, systems, and people working together such that many things that were previously done in painstaking and laborious ways, can now be done simply and quickly. Before the surveyor can fully embrace the technology, and become part of the process, the surveyor must understand that GIS means much more than "Get It Surveyed". It is my hope that the land surveyor will be more directly involved in GIS to the surveyor's benefit and in ways that enhance the value of GIS for everyone.

Some of the material in this book is compiled from articles that I have written and published over the past couple decades on the subject of Geographic Information Systems as it relates to the work of land surveying. I revised and updated some of the articles to reflect changes in the technology or the approaches to things. Other material is original for this manuscript.

Rj Zimmer, PLS

2013

HOW TO USE THIS BOOK.

This book is organized into seven major parts. The first part is an introduction to the origin and history of GIS, some general concepts, and a discussion of the supporting technologies followed by discussion of the relevance of GIS to land surveying and land surveyors.

The second part provides more technical discussions of key GIS concepts that are universally applicable. The intent of this part is to lay a technical foundation for using, understanding, and developing GIS technologies.

The third part covers GIS data sources – where existing data may be found and how to access those data.

The fourth part covers GIS use in the survey office that support surveying activities and project management.

The fifth part examines field applications of GIS in support of surveying, and field surveying in support of GIS.

The sixth part is devoted to spatial accuracy – a topic of special interest to the land surveying profession.

The final part examines a variety of topics on the relationship between the profession of land surveying and the types of activities that modern technology enables non-surveyors to engage in and the impact of technology on the definition of the practice of land surveying as a profession.

This book may be read non-sequentially. The reader may directly jump to most any section without having to read prior chapters. Each chapter is roughly self-contained. To achieve that independence, some content may repeat in order to provide relevant context, while major topics may refer to other sections of the book.

Additional training resources, including video tutorials, are available on the website http://GISforSurveyors.com.

Please send comments and/or corrections to RjZimmer@GISforSurveyors.com

PART 1 ABOUT GIS

This part is an introduction to an overview of Geographic Information Systems in general, and how GIS relates land surveying and the land surveying profession. The chapters of Part 1 are intended to provide context at a high level so that the reader can conceptually relate the subsequent more specific content. We present an overview of GIS databases, how they become spatial, why the spatial component is so powerful.

CHAPTER 1 HISTORY AND GENERAL OVERVIEW OF GIS

INTRODUCTION

Geographic Information Systems are a powerful combination of data, processes, computer resources, and people that provide a means to visualize and analyze spatial context and relationships. People use GIS for many purposes such as online searches for retail sources of products and services; navigational aids in cars, planes, and boats; for hiking and hunting; for health care services, and for a myriad of industries. Many GIS software are now easy to use and the supporting data are readily available with sufficient standardization to be useable for many applications. In fact GIS software that works in a web browser, or vehicle navigation products or on an iPod (GIS in the palm of your hand!), are so simple to use that people may not even realize that they are using GIS.

Because location is what surveying is all about, GIS is a great tool for land surveyors as it provides easy access to information in a spatial context. Surveyors use GIS in the office to manage survey databases such as control data, project locations and other information related to project sites. GIS also helps land surveyors with estimating potential projects by providing access to terrain information (such as digital terrain models and USGS topographical maps), aerial photography, land ownership, roadway access, institutional constraints (such as zoning, floodways, wetlands, etc.) and the built environment. Additionally many public agencies provide online access to survey records using a GIS mapping interface as an aid to performing records research. All this information comes together in a single graphical map interface, typically along with some simple tools for performing queries against the GIS databases that may be associated with some or all of the GIS features.

GIS has its origins in the 1960's as an aid to planning and resource managers who want to use computers to bring together various bits of information that share a common location. The initial concept was to use computers to represent map overlays of various themes which at that time were being done with linen, paper and polyester films. The computer allowed one to store, edit, analyze, and share the map overlays more quickly and easily. But GIS provided more than simple cartography because the data contained attribute information associated with the graphic information, thereby one to perform calculations for various analyses. During the early stages of GIS there was very little data in digital form and the computing tools where somewhat cumbersome to use. However, as computer technology advanced and the software became easier to use, more data were converted or collected, and more people were able to use GIS software.

GIS DATA SOURCES

GIS data come from many sources and it is now fairly straightforward to integrate one's own GIS data with data from other sources for purposes of mapping and analysis. In some countries national or federal programs create, update, and maintain GIS datasets with nationwide coverage, which are public resources, although in some nations, national datasets are not shared with the public or private industry. In those nations that do share their data; industry, the public, and all levels of governments use these federal datasets to perform their own work or to provide services for others. GIS is a robust tool that aids our understanding of our complex world.

Some examples of federal GIS data in the United States are the National Agriculture Imagery Program (NAIP) that provides spatially referenced aerial photography nationwide; the US Census Bureau's population data that get a lot of use by governments, non-profits, and industry for many purposes; the US Department of Agriculture's soils data and watershed data, and many more. Some private industries enhance some of the basic public datasets for resale by improving the scale, geometry and associated data, and, in some cases, packaging the enhanced data with software that simplifies access and use. In addition to federal datasets, state and local agencies also create and maintain GIS data for their internal purposes which increasingly also has value to the public. When the data are collected properly and made available they become a community resource which may foster economic development and enhance citizen participation. Supplementing these public datasets are commercial datasets which businesses create or enhance and repurpose for sale, or in some cases distribute freely (e.g. Google Earth). GIS users, such as land surveyors, may consume these data in a variety of ways. Today, one may physically transfer data via hard drives or disks, or download data through the internet or email, or connect to an internet map service to connect directly to distant GIS data. Additionally one may use an internet browser client application that consumes GIS content and provides tools to perform GIS analysis or business functions (Service Oriented Architecture - SOA).

Depending on our task, these existing sources may contain all the information that we need or not. For instance, the aerial imagery in Google Earth (a combination of aerial photography and satellite imagery from a variety of sources) may show us everything that we want to know about physical access to a boundary corner in a forest - where the roads are, the terrain, the vegetation. However, we can also use the existing data as a basis to create our own data and as context for overlaying our data. For example, we might use a county's parcel data and high-resolution aerial photography as a spatial reference upon which we build an index map of our company survey projects.

GIS POWER

The power of GIS derives from three fundamental components of a spatial dataset: a graphic, a location, and a database. These three components may define who, where, what, when, why, and how of each item of interest and of a collection of items.

The graphic helps us to see what something is. Because our brains process symbolic information very rapidly, GIS presents an answer to the question of "what" something is, by using representational geometry. The GIS graphic also helps us to see *where* something is, and it does so more quickly than if we were merely looking at a table of text and numbers or coordinates. The location information makes the data special because where something is located is as important as what it is. Location means a lot to people, as surveyors well know. Additionally, the juxtaposition of one feature to another based on their respective locations allows for spatial analysis. The types of spatial analysis may consist of counts of things: such as the number of control points within a county, or intersections of things: such as the total acreage of spotted owl home range that might be disturbed by a highway re-alignment project's clearing limits, the optimization of tasks: such as calculating the shortest route between two points along a pipeline network, finding and counting the number of nearest neighbors: such as how many control points are within one-quarter mile of a project area, as well as many other location based analyses.

Location information in GIS can take a few different forms. Coordinates are one of the most common means to locate or place a GIS feature. A point has a single coordinate set, but more complex geometries like lines, polygons, and raster data (such as digital elevation models, and photography) have multiple sets of coordinates that define their location and their geometry. Each GIS software program specifies the allowable datum, coordinate system, projections, and units that it will allow. Some software can re-project coordinates on the fly into a different projection, datum, coordinate system or units, to facilitate overlaying onto a single map multiple datasets that have different spatial references.

Although all GIS data are coordinate based, GIS software can use other types of location information to map new data to the proper place, the best example of which is street addresses. With special software, a process called geocoding relies on an existing reference dataset that is already in GIS coordinate space and that has road names and address ranges. The reference dataset is the basis for locating a new dataset by calculating points along a road where street addresses belong. Geocoding is a means to map many legacy databases into GIS.

DATABASES

GIS is very powerful because it is more than just a pretty map. GIS provides linkages between the graphics to tabular data. Figure 1 shows the various types of vector geometries of GIS graphics and a table of attributes which could be associated with those graphics. These linkages make it possible to add tables of information to each feature in a spatial dataset. The database part of a GIS dataset is the tabular information, which might be numbers, or text, or hyperlink strings to other information such as a scanned document. GIS data may have any number of associated information, including links to other databases.

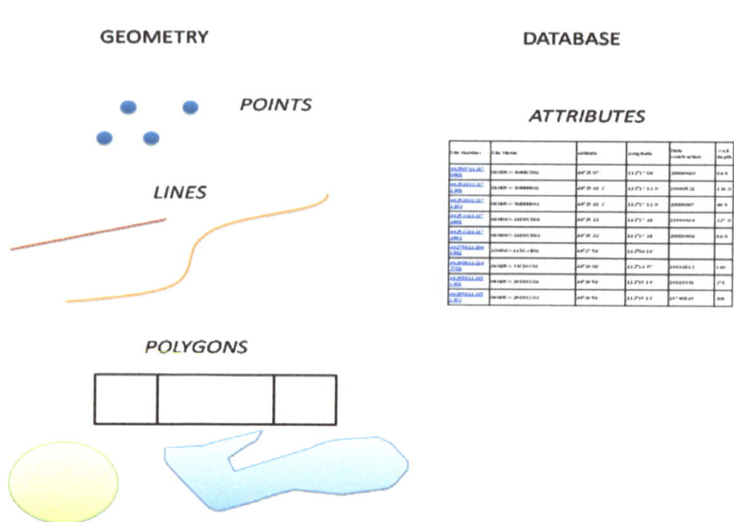

FIGURE 1 GIS GEOMETRIES AND TABULAR DATA.

The ability to access those databases allows one to select features that have particular characteristics. For example (see figure 2 Example Database Query), a database of control points may have information on when, where, and how the control was established, the coordinates' accuracy, the type of monument and other things. A GIS dataset of waterlines may have an associated database of pipe diameters, materials, and flow measurements. One can perform sophisticated queries and analysis against the database and have the results of those queries and analysis visually displayed to provide more meaning and context.

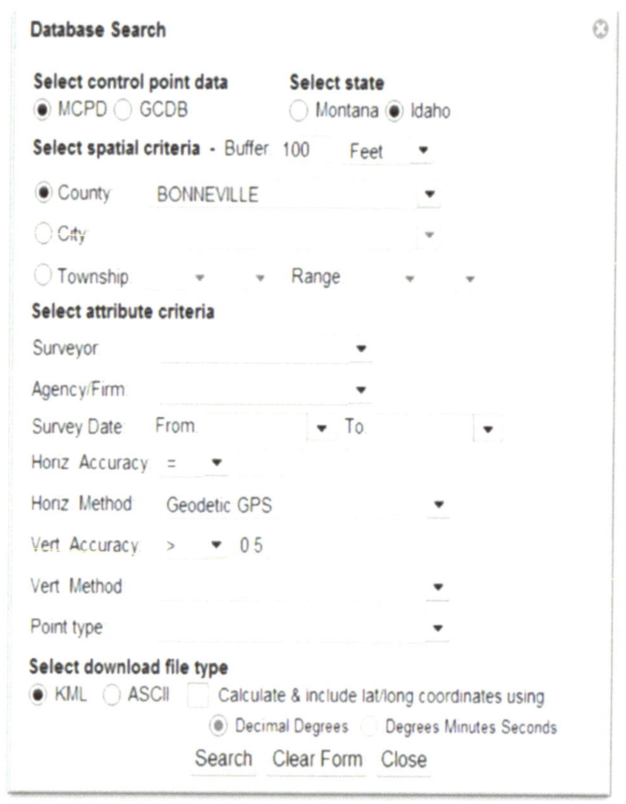

FIGURE 2 EXAMPLE DATABASE QUERY

The tabular data for a GIS dataset may already exist in some type of database, or one may need to convert existing data from some non-digital form such as paper records or one may need to create the data from scratch.

CREATING THE GRAPHICAL DATA

The most common methods for converting hard copy records into digital form are scanning and manual data entry. One can convert hard copy information directly into a GIS database or into an intermediate data format such as a spreadsheet, which one can then connect to the GIS data in a separate step. The preferred method for creating GIS data from *new* information is with GPS data collectors. GPS data collection automatically combines information gathering in the field with coordinate information derived from the GPS satellites. With the right equipment and field collection software one may also enforce good data structure and consistent data content.

GIS provides access to location information together with tabular data and that is a powerful combination. GIS is proliferating and nearly ubiquitous. There are many ways for surveyors to engage in GIS as part of our daily routine, with many opportunities available for surveyors to contribute to the fabric of GIS and spatial data.

CHAPTER 2 GIS VISION

Whether building an enterprise-wide GIS or GIS for a single user, it is important to understand and articulate the vision for that system. Typically one of the first steps to building a GIS is to discover the needs of the users, what deficiencies there may be in the current flow and use of information, and, perhaps most importantly, how the users envision things working in an ideal way. The idealized scenario should translate into the vision for the GIS. The vision for GIS describes the desired state of information flow and use. The vision articulates the state that the agency or individual desires to achieve, when the GIS is fully implemented.

For example, many institutions struggle with the inefficiencies of hard copy paper workflow. When many of the documents that one needs to use to do their work are only available in paper form then accessing those documents can be time consuming and inefficient. Furthermore, many of those documents may have originated from different sources, for a variety of purposes, which may make it difficult to relate them to each other when necessary. For instance, the US Federal Flood Insurance Rate Maps, which show special flood hazard areas and flood risk zones, may have been mapped at a scale different from an assessor map of the same area. The differences in scale of the two maps can make it difficult to identify the floodplain for a particular property. The vision for GIS in this example may be to have all the maps in a digital format for easy access, and be able to overlay those maps for easy use in order to be able to readily identify the floodplain for any property.

Before beginning to build a Geographic Information System, it is helpful to assess the needs so that the GIS can be built to meet those needs. When performing a GIS needs assessment it is essential to describe some of the goals for information access and use. It is also important to understand how to measure a successful implementation of GIS. One can then formulate those stated goals and measures of success into benchmarks for the implementation of a GIS. The benchmarks typically related to a current process which needs improvement, or it may be a task or process that cannot be done because of issues of data access or use. The benchmarks serve as checkpoints during the development of a GIS to guide the implementation. Successfully hitting those milestones during the GIS implementation will highlight the success of the GIS in ways that are clearly demonstrable to the potential users. Listed below are some goals expressed by GIS users:

- Have a road map that depicts the location of all roads in the county with road names, and other reference features.
- Have a set of road right-of-way maps that depict right-of-way widths for county roads, road names, jurisdiction, and ownership.

- Have on-line access to deeds, certificates of survey, plats and other documents of record in the records office
- Have access to information from the desktop
- Have connectivity to other departments & agencies
- Be able to create maps of fire hydrant locations
- Have desktop access to assessor rolls, legal descriptions, ownership, geo-codes.
- Be able to track special conditions on a property (land use compliance etc.) & tie it to the permitting process
- Be able to create a map that show public lands
- Be able to create a map that shows parklands, infrastructure, & park facilities.
- Be able to generate a map that shows streets and water or sewer line valves to shut off.
- Be able to create a map for one-call locates of sewer and water utilities.

These goals are things that people would like to be able to do, and are tangible, realistic things that GIS can help to achieve. Taken together the benchmarks can describe a vision. When working with an agency with a variety of disciplines, it is common for each discipline to have different needs, and therefore each may describe a different GIS vision.

Here are some typical examples of vision statements taken from various departments from different agencies:

A county public works department:

Public works envisions information on road rights of way, infrastructure, road location and resource use that is easy to access and easy to use. Public works would like to have the information it needs for improved decision making, when out on the roads and when in the office performing planning functions. Resource usage should be tracked in an automated fashion and tied to the road location where it is used.

A clerk and recorder's office:

These departments envision information that is easy to access, accurate, easy to use, and those operations perform efficiently. The records department envisions providing on-line digital access to recorded documents with an easy to use indexing and retrieval system.

A planning office:

The planning department envisions access from the desktop to all the data needed to support decision-making. The datasets should be fully integrated, spatially accurate and the information should be correct.

It is readily apparent that there are some common threads here. In the case of these particular departments' vision statements, the common threads are *easy access* to accurate and correct information; and that the information should be *easy to use*. By understanding the common elements of each discipline's needs and desires, one can forge a holistic vision statement that represents the entire enterprise's vision as a whole.

Generally speaking, people want the information that they need to be freely available, easy to find, and integrates well with other data. Additionally, full and complete metadata should be available for each piece of information so that one can evaluate its suitability. GIS can work for surveyors in a variety of ways that make it easier to perform certain surveying tasks. In later chapters we will look at specific examples of some of these tasks made easier with GIS.

CHAPTER 3 THE SURVEYOR'S ROLE IN GIS

Some surveyors contend that geographic information systems are deficient unless a professional land surveyor directs the data development and operations. The main issues tend to revolve around who may make digital representations of information that originated from surveying work. Some surveyors feel that only a registered land surveyor may legally create a digital representation of parcel maps, subdivisions, surveys and other boundaries that are parcel dependent (for example, taxing districts or zoning boundaries). Some surveyors contend that property rights, taxing concerns and public health, safety and welfare are at stake when survey and parcel data are used in GIS and, therefore, only a registered land surveyor could ensure the proper spatial accuracy necessary to create parcels and other boundary layers that are used in GIS. I have heard a surveyor state that all GIS data was "crap" if registered land surveyor didn't create it because of the accuracy issue.

The GIS community, for the most part, does not agree with this perspective, primarily for two reasons. The first reason is that the goal for GIS is not to create legal records but to create a visual index to legal records. Assessor maps are the best example of this. For decades, assessor maps have been hand drafted by cartographers who are not registered land surveyors. The purpose of the assessor map is to show who owns what and where that is. In order to assess how much taxes are owed, the assessor can draft a map that shows a parcel's approximate size and shape based on a deed or plat, and in which neighborhood the parcel is located. They do this by bringing together parcel configuration data from a variety of sources—deeds, surveys, subdivisions, court orders and so forth. The cartographer must graphically reconcile conflicting calls and take cartographic liberties (force a bearing or a distance) to complete parcel closures on the map. The map helps to show each parcel with the entire cadastral fabric, but the important assessment information is contained in tabular databases. This practice has been acceptable because these maps are for tax purposes only. The GIS community contends that a digital representation of a hand-drafted map should not necessarily be held to a higher standard. Therefore, because a license is not required to create the hand-drafted map, a license should not be required to create the digital version.

Assessor maps are converted to a digital format to make them easier to update. And while the parcel maps are created for assessment purposes, the digital form also provides better access for anyone interested in the data related to the parcels, such as ownership, assessed value and tax code.

GIS is a more efficient and convenient way to access parcel information. The real power of GIS, though, is that it makes it easier to overlay other datasets, such as roads, streams, floodplains and zoning boundaries with the parcel data. Early in GIS development many surveyors failed to understand and appreciate this power. However after using cadastral GIS to help with surveying activities (described later in this book), surveyors have come to understand the value of GIS even when surveyors have not done the GIS data development work. This does not mean that GIS data developed without the help of surveyors cannot be improved. In later chapters we will examine how surveyors can make a good product even better and thereby enhance the usefulness of the information.

The surveying profession changed a lot in the later part of the twentieth century as technological innovations put measurement tools into the hands of many. These technologies enable those with lesser skills to measure and map things that previously only trained surveyors could do. Those (within and outside the surveying profession) who think that measurement is the essence of surveying may see this trend as the beginning of the end of the surveying profession, but it is not. Instead of spending most of one's time measuring the mundane and the trivial, surveyors will refocus their attention to activities that are more professional in nature, such as analysis of measurement data and spatial relationships, and, the ever important land boundary work. As measurement and mapping activities become more accessible to everyone, the need for a surveyor to perform many kinds of measurements decreases, while the need increases for a learned and trained professional to evaluate and certify measurement data, and to provide context and meaning to those spatial relationships that measurements reveal. Geographic Information Systems and the new measurement technologies do not make the surveyor obsolete. On the contrary, these technologies reemphasize the need for expert survey analysis and evaluation of measurement data. Within the GIS framework, there are many different types of activities, of which measuring is one. The land surveyor performs all these activities to varying degrees.

In his book Exploring Geographic Information Systems, Nicholas Chrisman, a professor of geography at the University of Washington in Seattle, describes a GIS as an organized activity by which people:

- *Measure aspects of geographic phenomena and processes;*
- *Represent these measurements, usually in the form of a computer database, to emphasize spatial themes, entities, and relationships;*
- *Operate upon these representations to produce more measurements and to discover new relationships by integrating disparate sources; and*
- *Transform these representations to conform to other frameworks of entities and relationships. [Chrisman, 1997]*

Although often viewed from outside the profession as working only in the area of measurement, the surveyor works increasingly in many GIS activities. Many people might not identify some of the surveying activities as GIS activities, especially when the surveyor performs them outside a GIS environment. However, as GIS technology advances, the surveyor can perform more work directly within and in concert with the GIS environment. Chrisman also alludes to Data Quality: *Verify against World activities, that is ground-truthing measurements, and the quality assessment (including the spatial accuracy assessment) of measurement data* [Chrisman, 1997]. He also states that the measurements should support the goals of the project, and that the project should support the expectations of the institutions and cultures that demand the information. The land surveyor does have the experience, training, and knowledge to make the kinds of assessments necessary to determine the spatial quality of measurement data, and, the context and suitability of certain types of information (such as boundaries).

MEASURE

People measure phenomena in order to communicate their location to others. Measurement is a means to an end: that end being the communication of location of some thing or things. Whether the data are tied to a National Spatial Reference System, property corners, or other context, the measurements are always relative to other phenomena, and in this way, the measurer provides context for the locations. Depending on the nature of the phenomena, the measurements may be performed on the complete dataset (such as control points) or on portions of (representative samples of) the data. Examples include random elevation point measurements that are made to generate a topographic map, or point measurements which are then connected in sequence to form lines that represent a roadway.

In addition to measuring the location of phenomena, we may also observe and record other characteristics of the phenomena. For instance, we might perform a GPS control survey of a survey monument. When we perform that survey, we may also record characteristics of the control monument such as its material, markings, location, stability, etc. Typically, in a GIS, these observations are more important than the location measurements. Often within a GIS, the *what* takes higher precedence than the *where*. These observations may be recorded in the GIS as attributes to the data which can then be used as the basis for GIS analysis.

REPRESENT

Representation of measurements has become increasingly, though not yet exclusively, digital. In fact for the surveyor, the data may be digital from field to finish. However hard copy output, of plats and survey information continue to be important representations. In the GIS realm, we are beginning to see survey measurements represented in their raw form (Lidar data for example) in addition to derivative products such as elevation models. This trend is an important advancement for the integration of true survey data.

OPERATE

As Chrisman defines operate, more measurements may be made and new relationships between disparate information may be discovered, once data are brought together within a GIS. Surveyors perform important analysis of spatial data. Surveyors analyze measurement accuracy and analyze boundary conflicts for instance, but other types of analysis can be done using GIS. Surveyors can also perform site analysis and project planning using GIS topographic surveys, geology overlays, zoning and regulatory overlays and other relevant GIS data. Surveyors can also analyze the success and geographic distribution of their own projects in order to identify project profitability or to manage and allocate staff and equipment.

Chrisman contends that we measure in order to meet certain goals and those goals predicate the choices we make about what to measure and how to represent the phenomena. In turn, the availability of information shapes social and the cultural expectations of professions, while disciplines shape the choices of measurement and representation. The land surveyor, through experience and training, knows how to plan a project to make the correct choices about what to measure and the best means to make those measurements. Nevertheless, there is no way to predict what *new* expectations may evolve due to the availability of certain data. Misusing surveying data is one of the surveyors' greater fears with regard to GIS, but this is an area that no one has control over. The only caution that surveyors or anyone else can provide is to *clearly state in the metadata*, the intended and appropriate uses for the data. What happens after the data becomes available is hard to predict and nearly impossible to control.

The expectations of engineers and surveyors are different in many cases from the expectations of other professions and disciplines. Those expectations derive from the types of usage of the data. After all, we do not measure merely for measurement's sake; we measure in order to be able to do something else with that information. We measure for a purpose, in order to fulfill a goal. However, others who have access to the survey or engineering data may view the data in a different way and may see opportunities for other uses of the data. The classic example of this is the use of cadastral data. In some places, land surveyors feel proprietary about cadastral data and felt that surveyors were the only ones who could work with cadastral data. But they fail to understand that the cadastral *surveying* activities were not the entirety of a cadaster. A cadaster is more than mere boundary lines. A cadaster is a registry of land and real estate ownership. Property boundaries are one aspect of that registry. While the origin of a cadaster was for taxation purposes, today a cadaster has innumerable uses and applications due to the wealth of information contained in a modern cadaster. Thus, social expectations changed over time as data became available. The surveyor continues to have a role in cadasters (digital and otherwise), and performs many types of operations, including analysis of measurement data that fall within the GIS framework.

TRANSFORM

Transforming data representations to other frameworks (spatial and non-spatial) and into other types of relationships is also an activity performed by surveyors. Surveyors often transform legal descriptions contained in a deed or other record, into maps and into various digital forms. Surveyors also transform datasets from one coordinate system or datum to another. These types of transformations may seem trivial to the surveyor, but they are important activities in a GIS.

The surveying profession is evolving. While many people within and outside the profession see surveying as a profession of measurers, measurement is only one aspect of the profession; surveyors also represent, operate and transform information. Surveyors perform all the activities within the spectrum of the GIS framework. In doing so, surveyors are transforming as well.

CHAPTER 4 GIS AND THE SURVEYOR'S BOTTOM LINE

GIS can benefit the surveyor's business operations. Because surveyors depend on existing spatial information to do their work, improved access to and use of those data leads to time savings for information gathering, analysis, interpretation, and communication. Additionally, having more information available through GIS results in better decisions made in the office and in the field. Today ever more GIS datasets are available and GIS technology is increasingly easier to use. Therefore, surveyors who incorporate GIS into their work flow are saving time and saving money, and can work more effectively.

GIS helps the surveyor to locate and use information about places. Information such as land ownership, terrain, rights of access, monumentation, survey control, elevations, waterways and zoning, are now available for many places. The availability of useable information, particularly when the information is freely available online, saves enormous amounts of time for the land surveyor. The time required to research public records for land owner names and addresses, plats, aerial photography, zoning, existing infrastructure, etc. can been reduced from hours (or days), to mere minutes, in areas where federal, state, and local governments have fully matured geographic information systems. Additionally, copies of datasets on CD, online search, retrieval, download, and live connections to public datasets, provide surveyors with opportunities to incorporate GIS data into survey projects for analysis, reports, and communication. By incorporating GIS data in project proposals, surveyors demonstrate to potential clients, the surveyor's ability to integrate survey data with other client data. More and more clients are expecting surveyors to do just that.

Whether the client expects a GIS deliverable from the surveyor or not, the surveyor does save time by using GIS to acquire data for projects. By way of example, let's estimate possible savings for a small project's research. If we compare the estimated time required searching for and acquiring copies of public information without GIS versus the required when using GIS, we would see time savings for the research portion alone. In order for a surveyor to research and obtain copies of surveys, plats, and related public information for a project without GIS, time must be spent to drive to the county courthouse, to search by hand through a list of records, to make copies of the selected records, to pay copy fees, etc. Depending on circumstances, this work could take a couple hours out of a day, or even a couple days.

However, if those records are available online, and searchable through a GIS interface, and downloadable, or printable from the internet, then the time required to search and retrieve the information could be reduced to perhaps as little as 20 minutes. For a single research event, the savings can be an hour or two. Additionally, if at a later date or time, an additional document or

ACTIVITY	TIME WITHOUT GIS (HOURS)	TIME WITH GIS (MINS)	TIME SAVED	RATE	SAVINGS
DRIVE TIME TO AND FROM COURTHOUSE	0.75	0.0	0.75	$80.00	$60.00
PLAT LOOK UP	0.33	10.0	0.17	$80.00	$13.33
LAND OWNERSHIP LOOK UP	0.50	3.0	0.45	$80.00	$36.00
CORNER RECORD SEARCH	0.50	5.0	0.42	$80.00	$33.33
PROJECT TOTAL (HOURS)	2.25	0.3	1.8	$80.00	$142.67
YEARLY TOTAL (6 PROJECTS/MONTH)	162.0	21.6	140.4	$80.00	$11,232.00

documents must be obtained, they could be readily retrieved, instead of requiring an additional trip to the courthouse. Over the course of a year, even small savings like this can accumulate to satisfying savings in time and money, so that, even for a small company, using GIS can be profitable.

The benefits of GIS access to data can also be realized within a survey organization that has "spatially-enabled" its own data. A survey or engineering company or agency can use the advantages of GIS to organize and manage data for internal use. Company data such as client data, project locations, surveys that the firm has performed, corners that a surveyor has set, control point data, digital terrain models and other information that are mapped, stored, and indexed in GIS, can be readily retrieved and used. The ability to share information within an office, and between offices, as well as the ability to take copies data into the field, vastly increases the efficiency and the effectiveness of staff by providing ready access to the data used on projects. Communication among staff, as well as communication with others, in the office and in the field, is greatly improved, time is saved, and misunderstandings minimized, when all are sharing the same data.

With each passing day, as more GIS data become available and more data can be accessed online, in the office, and in the field, the importance of GIS increases. GIS today, is an essential tool for the land surveyor. GIS helps the surveyor to make informed decisions, to communicate effectively, and to save time and money. In later chapters we describe specific examples to illustrate these advantages.

PART 2 GETTING STARTED WITH GIS FUNDAMENTALS

This part introduces some basic GIS concepts that are essential to understanding what makes data spatial and how one GIS data set may relate to other data. An example data set, Digital Elevation Model, is described to illustrate a common GIS data set that is also important to surveyors. All GIS data come from somewhere and this part concludes with a discussion of some of the ways GIS data are created.

CHAPTER 5 GIS DATABASES & DATA MODELING

In this chapter we introduce the concepts of databases and data modeling for GIS.

DATABASES

A database is a collection of related data that are stored within a computer environment. A database may be comprised of a single file or multiple files which may be of the same file type or different file types. There are many types of databases and database management systems (DBMS). Data management systems are software programs written for creating and managing databases, and they typically provide methods for working with the data in the database such as creating records and loading data into the records, extracting data, editing the data, and methods for querying them. Not all DBMS support spatial data as categorically special. However, those that do differentiate the spatial properties, provide means to store the spatial information (such as coordinates, spatial relationships, topologies, etc.) and special methods for performing complex spatial queries. A database may be comprised of one or more tables. The tables may contain one or more records.

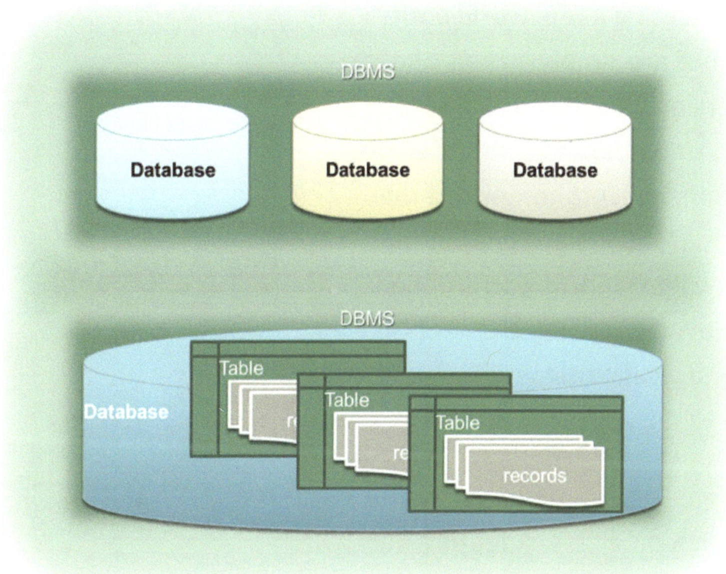

FIGURE 3 HEIRARCHY: DBMS – DB – TABLE - RECORDS

Tables collect a specified object of interest with a common set of attributes. For example, in a database table of Montana counties, the table would contain one tuple (row or record) per county and one column or attribute per property of each county (population, largest town, size, etc.). The set of tables contained within a database is called a schema. In GIS tables are associated with GIS graphical objects that are placed in geographic space. The tabular data support cartography and analysis with the GIS environment.

GIS DATA

GIS is a collection of computer hardware, software, and geographic data for capturing, managing, analyzing, and displaying all forms of geographically referenced information. A GIS database is a type of database which, in addition to storing graphics and attribute data, also stores coordinate information or some other type of spatial positioning information. GIS databases store data in themes, which are somewhat equivalent to tables in relational databases.

A theme is a GIS *layer* consisting of the spatial location and a common set of attributes representing a real world entity. Themes are simply groups of data. Layers may be viewed on-screen in a GIS or on paper as part of a map ((multiple layers comprise a map). In a map, layers are symbolized with colors and/or patterns and displayed with map elements like a legend, scale, title, etc.

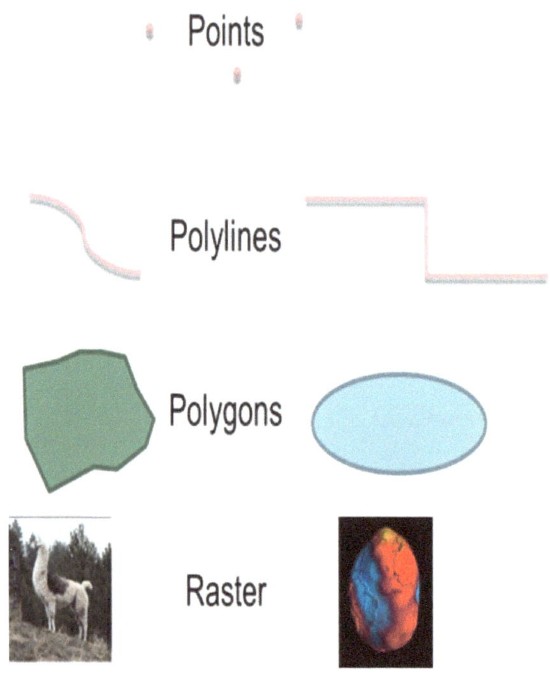

FIGURE 4 GIS OBJECTS

GIS data are used to model real world phenomena, or features, within a database. A feature may be a survey monument, or a road, a governmental boundary, a land cover, an elevation point, and any number of other things. Features are represented in a database by data objects. In GIS the shape of data objects may be represented by points, lines, polygons or grids (rasters) as in figure 4 . The collection of these objects and the relationships between them are defined by a data model as shown in figure 5.

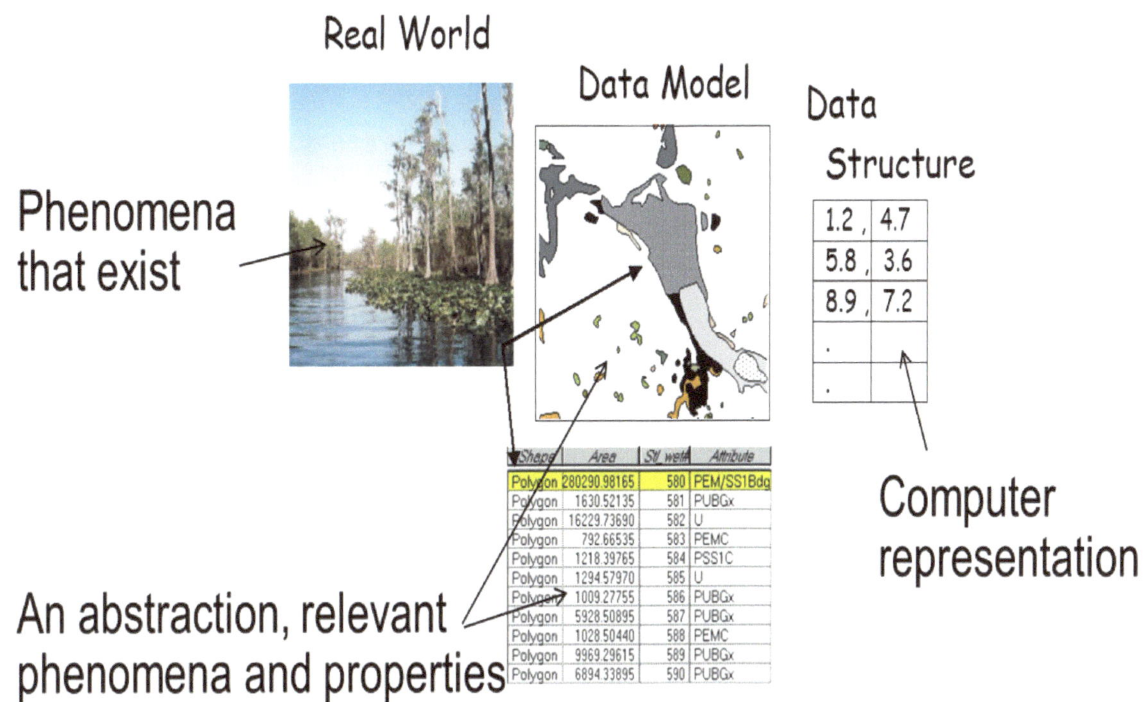

Shape	Area	Stl_wetA	Attribute
Polygon	280290.98165	580	PEM/SS1Bdg
Polygon	1630.52135	581	PUBGx
Polygon	16229.73690	582	U
Polygon	792.66535	583	PEMC
Polygon	1218.39765	584	PSS1C
Polygon	1294.57970	585	U
Polygon	1009.27755	586	PUBGx
Polygon	5928.50895	587	PUBGx
Polygon	1028.50440	588	PEMC
Polygon	9969.29615	589	PUBGx
Polygon	6894.33895	590	PUBGx

FIGURE 5 GIS DATA MODEL REAL WORLD INFORMATION

THE SPATIAL INFORMATION

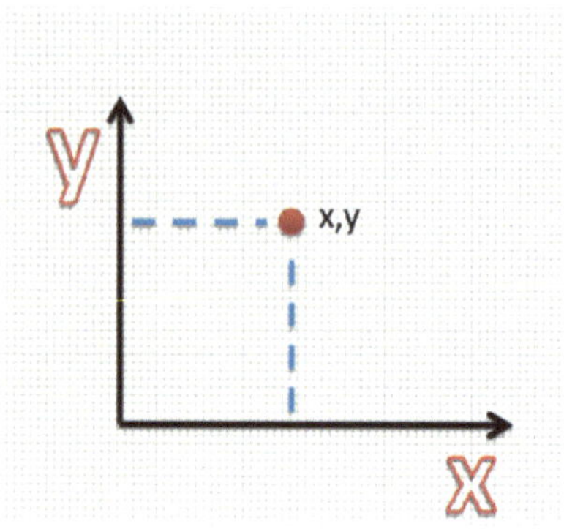

FIGURE 6 CARTESIAN COORDINATE SYSTEM

Location makes GIS data special. Location places a GIS feature in geographic space. Additionally, non-point features must contain information that describes the length or area occupied by the feature in that geographic space. The location of things can be expressed in many ways within GIS databases but normally the spatial characteristics are stored as numerical values of a Cartesian or spherical coordinate system (figure 6). A GIS feature may have a single coordinate, an ordered series of coordinates, or a single coordinate plus relative coordinates. The location of GIS features may also be expressed as a relationship to other existing GIS features – relative referencing, which typically are used to generate coordinates based on the reference data (see the chapter on geocoding).

Example location expressions:

LOCATION TYPE	LOCATION EXAMPLE	GIS DATA EXAMPLE
COORDINATE PAIR (NORTHING, EASTING):	-111.90736860291,46.7083937266146	POINT FEATURE (E.G. CONTROL POINT)
BEGINNING POINT, SCALE, SIZE:	NCOLS 7 NROWS 7 XLLCORNER -7589988.98 YLLCORNER 3995608.83 CELLSIZE 926.62543305583381	ASCII GRID (E.G. RASTER SOIL MAP)
COORDINATE STRING:	-111.978115064061,46.5908599277789,0 -111.975469276482,46.5908713357322,0 -111.97547006816,46.5907490233486,0 -111.974282561415,46.5907501339878,0 -111.974044979296,46.5908135751861,0 -111.9740558195,46.5911954435711,0 -111.974094222589,46.592010507787,0	VECTOR POLYGON (E.G. PARCEL BOUNDARY)
RELATIVE (ADDRESS)	123 N MAIN ST, GIS TOWN	XLS SPREADSHEET

DATA MODELS

A data model is a collection of data objects and relationships represented in a database. A data model describes a structure for representing data. The data model is determined by the level of abstraction of the data and should support storage, query, and analysis needs. The main types of data models are conceptual, logical, and physical (figure 7). These types embody the realm of the model.

The conceptual data model defines the entities and the relationships between them; independent of a specific implementation or system. The logical data model organizes the conceptual data model into components and relationships which are explicitly defined. The physical data model describes how the logical data model is structured in a DBMS tables and relationships.

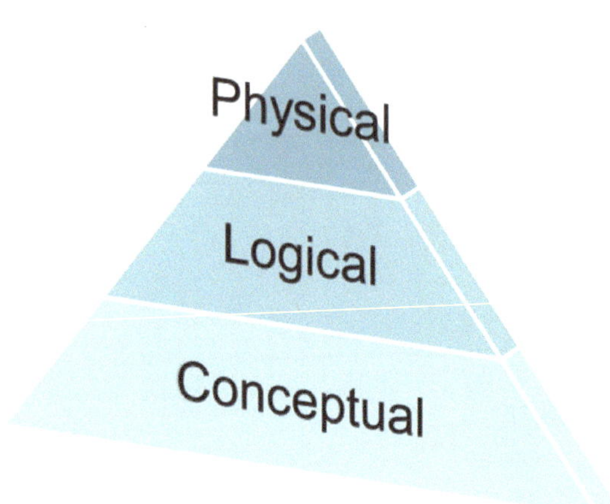

FIGURE 7 HIERARCHY OF DATA MODELING

In all cases digital data are representational. This is important in GIS because a data model is only a representation of a sampling of the real world. The challenge is how to best represent real world phenomena digitally.

Representation is based on perception and measurement, typically of a sampling of the phenomenon. Ideally, one wishes to accurately represent reality but digitally we cannot contain the actual thing – only partial representation of the thing.

For example, as in figure 8, a patch of forest (phenomenon) may be represented by a polygon (object) that represents the location and extents of the entire forest, or by a set of polygons that represent the various dominant tree species within the forest. These representations may work well for large stands of forest, but small stands (such as in a city park) may be modeled as individual trees.

FIGURE 8 MODELING A FOREST

Since data modeling is an abstraction of real-world phenomena which are complex and perhaps constantly changing, we make tradeoffs when trying to model things. We must consider that we may represent things differently at different scales. We must decide what amount of detail can be left out. We must identify our area of interest, we must set minimum scale for mapping the data, and we must choose a specific time for when our model is a snapshot.

In all cases the model should be tied to the purposes for which we shall use the data because what we model and how we represent it affects what we can do with it. Cartography, analysis, storage, maintenance, editing, updating, overlaying with other data and providing access, all have inherent requirements for how the data are modeled.

FILE FORMATS & STRUCTURES

There are a number of file formats and data structures that specifically support spatial data. Additionally, most GIS software will also support non-spatial (non-GIS) data as well although that support may include limitations on the ability to query or edit or analyze some formats. GIS data may be represented in many different formats for both vector data and raster data. File formats have evolved over time with the various updates of GIS software. However use of some formats may persist long after the software is obsolete. The ESRI shapefile format is an example of a persistent format for GIS vector data. Here we discuss a couple common formats for GIS data.

The two main classes of GIS data are, vector and raster. Many data may be represented in either format although one or the other may be more efficient for the types of operations one wishes to perform or the volume of data. In some circumstances the data may be converted from one to the other in order to perform certain operations more efficiently. Raster data is the typical format for imagery (such as aerial photography), soils, geology, digital elevation models (DEMs) and the types of data that cover very large areas. Vector data are normally used for more discrete (as opposed to continuous) features that may be intermittent on the landscape such as survey control monuments, roads, property boundaries. Raster data define an area covered by values. Vector data define objects at locations and also contain tabular data as attributes for each feature.

One common raster format is a GRID. The GRID format is typically used to store elevation or land use or land cover data. A common vector format is the ESRI shapefile which is a homogeneous collection of features that may represent GIS features of a single primitive type (points, multipoints, polylines, or polygons). The shapefile format is comprised of 3 essential files, one for the graphic (.shp), one for the tabular data (.dbf), and one file that contains linkages between the first two files (.shx). There may be other files of the same name but with different 3 character extensions. These extra files are helpful but not essential. Most GIS and Computer Aided Design software support the shapefile format for viewing and display, and some may allow editing or creating shapefile geometries or attributes.

GEODATABASE

Increasingly common, especially for complex data, is the geodatabase format. There are many brands of geodatabases, some proprietary and some open source. But the advantage of a geodatabase is that it provides a means to store large amounts of GIS data and data of different types. A geodatabase can simultaneously contain raster data and vector data that are spatial and non-spatial. Some geodatabases may support relationships such as topology (spatial relationships) and table relationships between datasets within the geodatabase. Geodatabases are supported by multiple file formats and software. Some geodatabase management systems support simultaneous multiple-user access and editing.

CHAPTER 6 SPATIAL REFERENCE SYSTEMS

GIS data contain coordinates which are referenced to a coordinate system, on a particular datum, in a geographic projection of some kind using some type of common measurement units. Taken together these specifications are referred to as the Spatial Reference System (SRS) of the GIS data.

For example, to create a location on Google Earth you must specify 3 numeric values: longitude (decimal degrees between -180 and 180), latitude (decimal degrees between -90 and 90), and altitude (meters) – example:

```
<Point>
  <coordinates>-90.86948943473118,48.25450093195546</coordinates>
</Point>
```

These are geographic coordinates (not projected) in the WGS84 datum. The units are decimal degrees for the horizontal coordinates and meters for the altitude.

6.1 PROJECTIONS

Map projections such as those shown in figure 9, are used to mathematically map three dimensional coordinates into a two dimensional plane. Thus projections will always distort the data in either areas, shapes, scale, angles, or distances. There are many different projections, each with its objective to minimize the distortion of one quality over another. Projections are independent of coordinate system used, although *plane* coordinate systems will be based on some type of

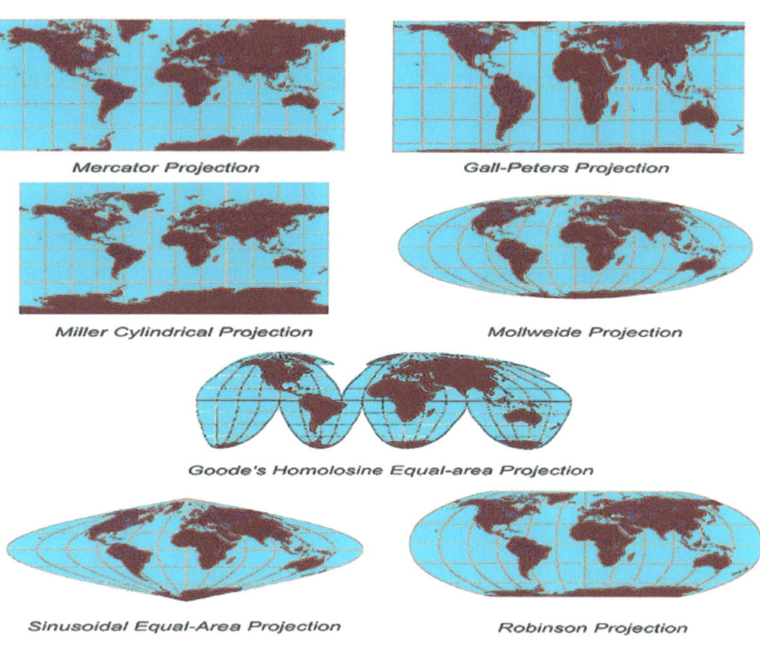

FIGURE 9 PROJECTIONS MAP 3D SPACE TO 2D SPACE. (IDAHO STATE UNIVERSITY, 2013)

projection.

6.2 COORDINATE SYSTEMS

A coordinate system defines a measurement space in two or three dimensions. A coordinate system may be in a projected space, or not. The coordinate system definition describes the origin, the direction of the axes, and the measure to use. The measure can be linear, curved, or angular. Most people have used simple coordinate systems to draft plots on graph paper. There are many types of coordinate systems in use in GIS. An example type of coordinate system commonly used in GIS is the State Plane Coordinate System of the United States. State Plane Coordinate Systems (SPCS) are flat earth, that is *projected*, coordinate systems that simplify measurements. There is a SPCS for each state. Because SPCS are projected, they introduce distortion in the measurements as a trade-off for the simplicity of the measures. Some states are divided into multiple state plane zones in order to reduce distortion.

6.3 DATUMS

A datum is a basis for measures within a horizontal or vertical coordinate system. A geodetic datum is a set of reference locations along with assigned values from which other things may be measured by reference.

For example, a vertical datum may have an origin on a rock somewhere in the world and the coordinate assigned to that origin may arbitrarily be designated as 1,000. Other things within that vertical space are then measure relative to that rock (location) and its value (vertical coordinate).

6.4 MAPPING UNITS

The mapping units are the individual measurement quanta. Typically in GIS, map units are angular units such as degree, or linear units such as feet or meters. Angular units may be expressed in various ways. Typical angular units are listed in the chart.

ANGULAR FORMAT	EXAMPLE
DEGREES, MINUTES, SECONDS	29° 37' 26"
DECIMAL DEGREES	29.62388°
DEGREES, DECIMAL MINUTES	29° 37.433'

All GIS data are referenced in a coordinate system with some type of spatial reference system. Each GIS data set can have only one SRS. Most GIS software has methods for assigning a spatial reference system (projection/coordinate system/ datum/ units) to a data set and/or a map (which is composed from GIS data sets). In order to overlay GIS data that have *different* spatial references, some, all, or many of the GIS data may have to be re-projected to a different spatial reference system. Some GIS software will do this *on-the-fly*, which means it can do this operation without changing the

SRS of the GIS data file – it merely calculates the re-projection then mathematically presents the data in the projected geographic space inside a map.

CHAPTER 7 TOPOLOGY

Topology is the branch of mathematics concerned with those properties of objects that do not change when the object's geometry is twisted, stretched or deformed. According to Merriam-Webster, topology is the branch of mathematics concerned with those properties of geometric configurations (as point sets) which are unaltered by elastic deformations (as a stretching or a twisting). *"Topology can be used to abstract the inherent connectivity of objects while ignoring their detailed form."* (Eric W. Weisstein, Encyclopedist ,Wolfram Research, Inc.). For our purposes, connectivity can exist between line segments, as in a pipeline or stream, or between polygons representing parcels or subdivisions.

For instance, the relationship between different segments of a stream can be described as a linear network. The topological rules of the stream describe the behavior of the segments with respect to each other and irrespective of the shape of the stream. Those rules might include things such as: each segment must connect to one or more segments in sequence, and the flow must go only in one direction. These topologic rules would apply regardless of any geometric alteration that the stream may undergo. That is, the segments must connect and the flow must go in only one direction whether the geometry of the stream is a straight line or a curved, sinuous line. Changing the shape of the stream does not change the topologic relationships.

Topology is germane to surveying where the concepts are applied frequently, although, perhaps not thought of in those terms. Topologic rules that apply to real property, for example, include such things as parcels must not have gaps or overlaps, and that a parcel's boundary must form a closed polygon. These topologic rules apply to parcels regardless of the shape of the parcel, and apply to a parcel even when its shape changes. These are important concepts which surveyors apply to boundary work. When a boundary line adjustment is made between lots, the topological relationships are maintained: the lots remain contiguous, they do not overlap, and no gaps exist between them.

Within GIS, topology is used to model data, or to model behaviors of GIS objects. GIS applies topology concepts to geometric objects in order to model the behavior of the collections of features in a dataset, and to describe the relationships between different datasets. Different types of GIS objects (polygons, lines, etc.) may have different topologic rules, as exhibited above. Additionally, GIS can even model topological relationships between different collections of objects. A subdivision polygon would have a topological relationship to the boundaries of the lots that comprise it, or voting precincts may have topological relationship with census block boundaries.

Topology rules can be enforced during the creation and editing of datasets to ensure the integrity of the relationships. Thus when parcel lines are edited for instance (such as for a boundary line adjustment), the topology rules help to avoid errors such as moving one parcel line without moving the adjacent parcel line thus avoiding gaps or overlaps as shown in figure 10. In addition, topology allows behavior modeling for geometric networks, polygon geometries and other types of GIS objects. The advantages to using topology come when analysis and queries are performed on the data. For example, if topology rules for linear networks are applied to a GIS representation of a sewer system, then the system's flows can analyzed.

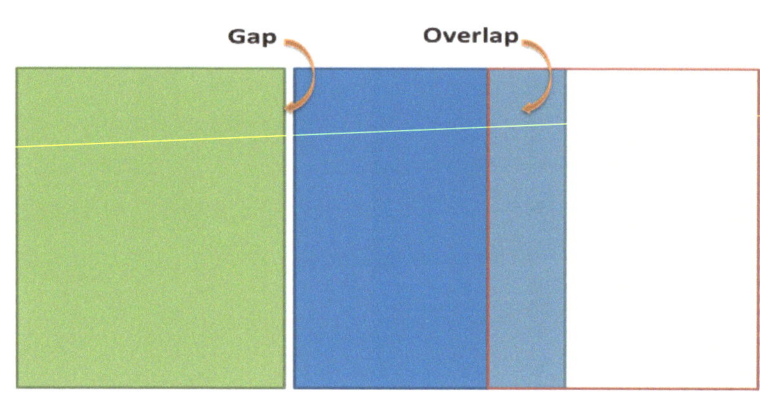

FIGURE 10 EXAMPLE TOPOLOGY RULE: PARCELS MUST NOT HAVE GAPS OR OVERLAPS

When developing a GIS or converting data into digital form for GIS, it is always important to understand the questions that the data will need to answer so that the design, including the topology rules, supports those questions. Just as there is no one way to build a truck, there is no one way to build a GIS dataset. The purpose for which the data are used must be fully understood at the outset, and then appropriate topology rules applied. While it may be true that parcels polygons must not have gaps or overlaps, a GIS dataset of weed polygons, might overlap and have gaps. When modeling linear networks, different topology rules might apply to water systems than to road networks.

Purpose is important when it comes to parcels representing property boundaries. Surveyors typically are called on to define the location of parcel edges and angle points, that is, the polygon *shape* is of primary importance. Because surveyors tend to concentrate on the boundary aspects of real property, there is a tendency among surveyors to cringe at anything that does not honor those explicit boundary geometries. Nevertheless, there are far more applications in GIS for which the boundary information is of secondary importance or even completely irrelevant. For instance, if one is only concerned with the value of land within a certain geographic area, or when querying a cadastral dataset to find the owner name and contact information, the shape of the parcels does not matter as much as the general location of the parcels. Thus, for many GIS applications, changing the boundary geometry does not affect the answers being asked of the data. The mathematics of the boundary, which is the bread and butter of the surveyor, is incidental to most GIS applications because the topologic rules are more important than the boundary geometry. That is why stretching or twisting of the geometry has little or no effect on the ability to perform the most of the important GIS functions.

The main difference between GIS and surveying, when it comes to topology is a matter of perspective, which stems from purpose. Surveyors are often focused on physically and mathematically, describing the edges of parcels, that is, the boundaries of objects. Whereas most users of parcel data in GIS are concerned with what applies within that configuration (what touches it, what is inside the parcel, what is the parcel inside of, what is near it, what crosses it) as well as some information extrinsic to the geometry (who owns it, how much is it worth, etc.). In instances where the surveyor is concerned with who owns a parcel, or what zone is it in, the parcel's edges are not necessarily important either. Yet, when the surveyor must monument the corners of the parcel, or plat a new parcel, then the edges are of primary importance.

PART 3 GIS DATA SOURCES

In the world today there are thousands of GIS datasets. While not everything has yet been mapped anywhere, there are some countries and some places in some countries where the GIS features sets are complete and rich for certain categories of data. These data were created by governments, non-profits, commercial interests, and individuals. Before beginning a GIS project it is always important to find out what data are already available. What you need may already exist, and if not, what you must create may depend on other data that already exist. In this chapter we will discuss Spatial Data Initiatives which many nations and states have or are working on, and the interdependence of GIS Layers which is important to understand when creating, updating, or using GIS data.

CHAPTER 8 GIS FRAMEWORK DATA & NATIONAL SPATIAL DATASETS

Agencies and people working together to build, maintain, update, and distribute GIS are the reason that some places enjoy rich GIS data resources. When many parties have an interest in spatial data which require large investments in time and money, then it more effective for those parties to pool their money, talent, and time to work in concert to build unified data that all may share. There are a number of datasets that most people using GIS need so these datasets are identified as *framework* data. Framework data are useful in themselves and they also provide a basis upon which to build other data. For example, in order to build any GIS data one must start with a network of survey control to provide the basis for assigning geographic coordinates to other data (for more detail on the interdependence of GIS layers see the next section). The coordinate data development, maintenance, and distribution efforts are called Spatial Data Initiatives (or SDI). These initiatives may vary in scope from municipal, to regional, to state/provincial to national to international to global depending upon the area covered and the participants involved. What they all have in common is coordinated data development, maintenance, and sharing. However, they vary in whether they share some or all of their data with the public.

THE SEVEN NSDI FRAMEWORK LAYERS:

1. GEODETIC CONTROL,
2. ORTHOIMAGERY,
3. ELEVATION,
4. TRANSPORTATION,
5. HYDROGRAPHY,
6. GOVERNMENTAL,
7. CADASTRAL INFORMATION.

In general SDIs are created to promote sharing geographic data. The United States Federal Geographic Data Committee (FGDC) has a national SDI (NSDI) that engages federal, state, and local partners in developing a set of eight framework layers which are shared among the partners and the public. The goals of the US NSDI is "... to reduce duplication of effort among agencies, improve quality and reduce costs related to geographic information, to make geographic data more accessible to the public, to increase the benefits of using available data, and to establish key partnerships with states, counties, cities, tribal nations, academia and the private sector to increase data availability" (FGDC 2013). The NSDI framework promotes multi-agency data development and sharing, by focusing efforts and finances on collaborative projects.

NSDI participation is voluntary, open, flexible and adaptive collaborations for shared capital planning, building, using and financing spatial data. They optimize and align the interdependencies allowing institutions and citizens to rely on and share quality data from other trusted sources.

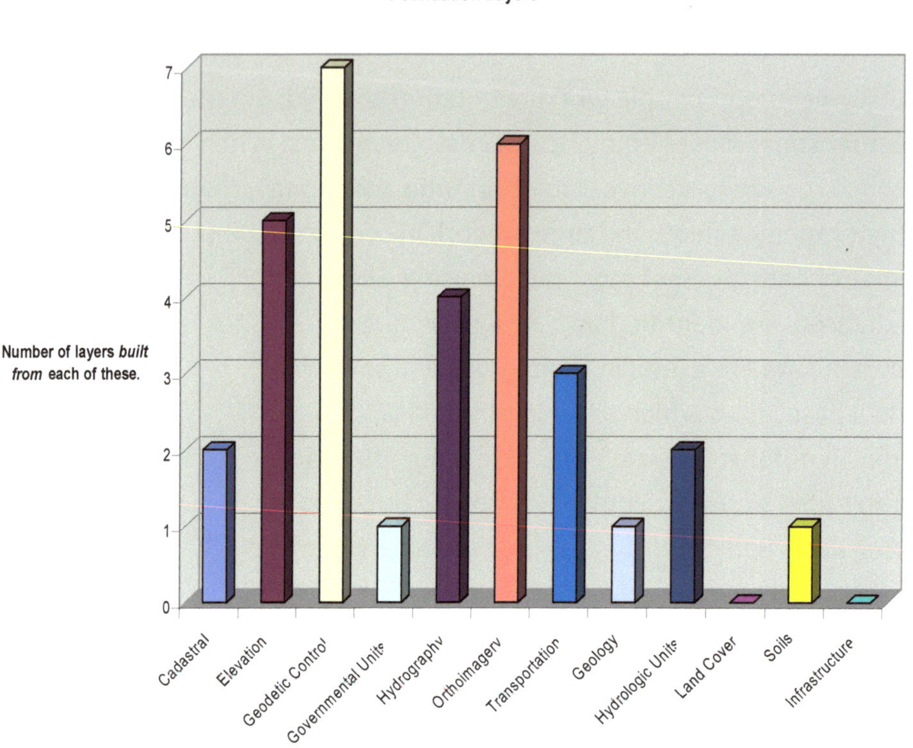

FIGURE 11 FOUNDATION LAYERS THAT SUPPORT FRAMEWORK LAYER DEVELOPMENT

The NSDI supports seven FGDC framework layers. Framework layers are those layers that have been identified as the most necessary and fundamental GIS layers. Some state participants expand that list beyond the seven framework layers. Figure 11 shows how one agency's GIS depends on the seven NSDI layers and a few other foundations layers, which, in turn, support the construction of many other GIS layers.

RELEVANCE OF SDI TO SURVEYING

There are two types of interest that surveyors have in the SDI framework layers. One interest is that of a data contributor. The other interest is that of a data user. Surveyors have been engaged in developing and updating some of this framework layer. Additionally, surveyors often use some these layers such as elevation data and aerial photography, in their work. Indeed, anyone that has used Google Earth has used framework data since Google uses NSDI orthoimagery in some areas.

SURVEYORS AS DATA CONTRIBUTORS

The surveyor is a contributor for most of the seven framework layers and an important contributor for some. Geodetic (or mapping) control, orthoimagery, digital elevation, transportation and cadastral data, are all typically within the realm and scope of the types of information that surveyors create. In some instances the surveyor may play a support role such as providing survey control for orthoimagery projects. In other instances, such as geodetic control, or digital elevation data the surveyor may be the only reliable source of the data. In either case the surveyor is an important player in the data development activities.

SURVEYORS AS DATA USERS

The framework layers are used in many public and commercial applications, such as the one shown in figure 12. From the user perspective, the surveyor is interested in every one of the framework layers and other non-framework layers. Every one of the framework layers is relevant to the surveyors work. The ability of the surveyor to access those data digitally can help the surveyor to work more effectively and efficiently if those data are developed in ways that meet the surveyor's needs. The surveyor needs to be involved with the development of framework layers to ensure that the results are products that work for the surveyor as well as the rest of the GIS community.

FIGURE 12 EXAMPLE NSDI LAYER – USGS IMAGERY IN GOOGLE EARTH

WORK OPPORTUNITIES

It's also important to keep in mind that work opportunities for surveyors may come out of these SDI plans. For instance , the US Bureau of Land Management uses survey control on Public Lands Survey System corners to improve the spatial accuracy of the BLM"S Geographic Coordinate Database upon which cadastral data are built. The resulting improvement in spatial accuracy of the cadastral data benefits all GIS users, for example, parcel lines will better align with aerial photography.

INTERDEPENDENCE OF GIS LAYERS

There are important relationships among GIS datasets that should be considered when

	Cadastral	Elevation	Geodetic Control	Governmental Units	Hydrography	Orthoimagery	Transportation	Geology	Hydrologic Units	Land Cover	Soils	Infrastructure	This layer needs
Cadastral			D		D	D	D						4
Elevation			D										1
Geodetic Control													0
Governmental Units	D	D	D		D		D						5
Hydrography		D				D							2
Orthoimagery		D	D										2
Transportation			D			D							2
Geology			D										1
Hydrologic Units		D			D								2
Land Cover					D	D							2
Soils						D							1
Infrastructure	D	D	D	D	D	D	D	D	D		D		10
Layer needed by	2	5	7	1	5	6	3	1	1	0	1	0	

developing or maintaining GIS data. Often times GIS data conversion efforts focus on converting a single layer into digital format without regard to the relationship one GIS layer may have to other GIS layers. For example, orthophotography requires geodetic control, so orthophotography projects depend on geodetic control.

The chart shown in figure 13 illustrates some interdependencies among GIS layered. There are two ways to view this chart and the two ways provide important information. The first way, reading down the chart to see which layers on the left need the layer on the top. Some layers of the layers on the top support many other layers; some do not support any other layers (as summarized by the counts at the bottom of the chart). The layers at the top that are most needed in order to build the layers on the left are the Foundation layers. Figure 14, shows the list of the Foundation layers and their relative importance to framework layer development. Geodetic Control plays a critical role in GIS data conversion for nearly every framework layer. There are only two layers (Soils and Land Cover) that can be developed somewhat independently of Geodetic Control, although fundamentally all positioning derives from geodetic control. In addition to Geodetic Control, Elevation data and Orthoimagery are the other two most important foundation layers. There is no framework layer that can be built without using at least one of these three layers.

The other information that can be seen from the chart is which layers are the Dependent Layers which can be seen by reading across the chart. These are the layers that depend on other layers for their creation. For example, in order to create Orthoimagery, Geodetic Control and Elevation data must be obtained. We can see from figure 14 that all framework layers are dependent upon other layers, with the exception of Geodetic Control. Geodetic Control is the only framework layer that can be built independently, although other framework layers can be helpful in planning survey control projects (as discussed later in this book). It is interesting to note that the layer termed "Infrastructure" depends upon nearly every other layer for its creation. In the SDI context, Infrastructure is a generic term denoting many of the features built by man, and includes bridges, dams, power generating facilities etc. One can see from the relationships in this chart that in order to develop the Infrastructure layer (critical or otherwise) many of the other layers must already be in place or developed first.

This information helps us to understand the importance of individual layers and the interdependence of the various framework themes. Thus, for example, in order for any layer to be created, Geodetic Control must exist. Also, the Cadastral layer is supported by Geodetic Control, Transportation, Orthoimagery and Hydrography. Any social, political or economic issues the use GIS applications for decision making, and thus demand one of the framework layers, will in most cases, also require one or more other framework layers. Therefore, GIS data development projects should take a holistic approach that supports the development and use of the foundation layers in addition to

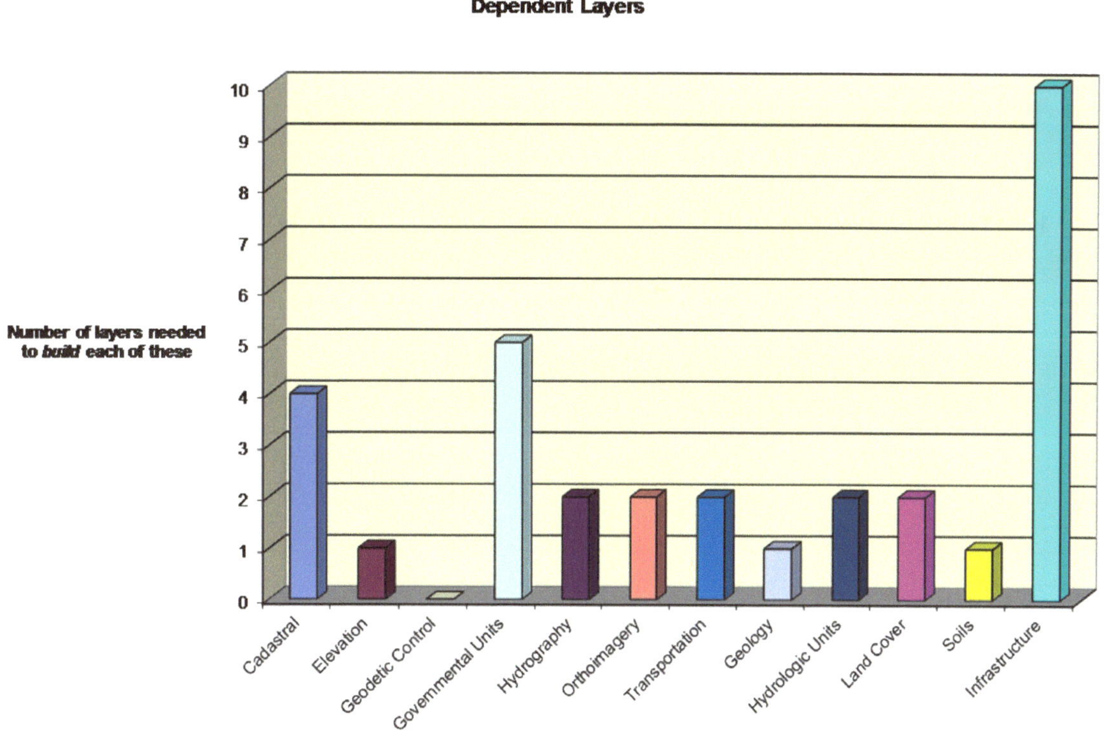

FIGURE 14 DEPENDENT FRAMEWORK LAYERS THAT SUPPORT THAT REQUIRE OTHER LAYERS FOR DEVELOPMENT

the layer of interest.

CHAPTER 9 CAD TO GIS

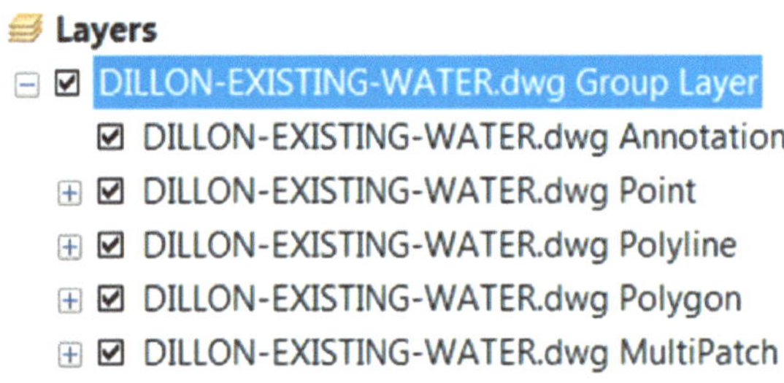

FIGURE 15 CAD DATA TYPES INTERPRETED IN GIS

Surveyors have been using Computer Aided Design/Drafting (CAD) for decades to draft surveys and to do design. As society becomes more spatially conscious, the ability to integrate data with geographic information systems data increases and is now to the point where CAD-GIS integration is expected by clients and is necessary for analysis and design of many survey projects. With the appropriate GIS software one may directly view CAD files and CAD features in a GIS environment in order to display a visual representation of the CAD elements in context with GIS data such as shown in figure 15. Alternatively, one can display GIS data in CAD software. However, in order to empower the CAD data for GIS spatial analysis, the CAD features should be converted to a GIS format that have feature attributes assigned to them.

Here we will look at the simple steps necessary to convert a typical CAD dataset to GIS format. Our example dataset represents the water lines of a small municipality. The water lines are polyline entities in an AutoCAD drawing which has a layer schema that defines the pipe diameters (a layer for each pipe size). The city has used the CAD as a basic inventory map to show what the city owns and where the pipes are located - purposes for which the CAD model adequately meets the needs. Now, however, the city would like to do a number of other tasks which can be more easily done with GIS such as, quickly query the database for inventory and reporting, take the data into field to navigate to manhole locations with GPS, edit and update features and attributes while in the field, perform connectivity analysis, be able to quickly and easily generate reports, create maps of the pipes overlain on aerial photography, parcel ownership, and utility maintenance districts, and be able to perform capacity analysis for build-out modeling. All these additional things are easily handled by GIS software, so the city must convert the CAD data to a GIS format.

DATA TYPES IN CAD VERSUS DATA TYPES IN GIS

All GIS data are representational, that is, the GIS graphical data represent the real world data in a symbolic manner. CAD data are also representational although the way of modeling the data differs from the GIS data model. CAD data generally represent large-scale objects while GIS data typically represent smaller scale objects. In our example project dataset, the CAD data consist of points, text, polylines, multipatches, and polygons. The features of the CAD dataset fall into one of these feature types and the features also have other characteristics, such as color and line type.

In Figure 16 (Attribute Properties of CAD Feature) shows the CAD characteristics as attributes of one of the water lines in the dataset (interpreted as a polyline in GIS). When we convert the CAD data to a GIS model, we may choose to retain those characteristics or not.

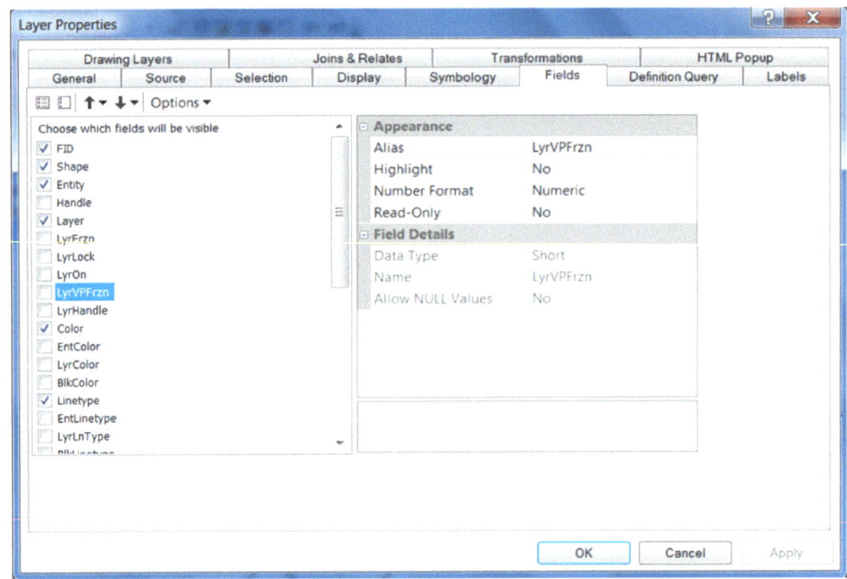

FIGURE 16 ATTRIBUTE PROPERTIES OF CAD FEATURE

When the CAD dataset is loaded into GIS, the features are organized into the appropriate feature type, and each feature's characteristics are represented as *attributes* of the feature. Figure 17 shows the CAD file loaded into GIS in a roughly geo-referenced location.

SPATIAL REFERENCE

Since GIS data are always contained within a defined geographic space, the CAD data must also conform to this requirement in order to align with other geographic data. If the CAD data are not already geo-referenced as in Figure 17 where the scale and location of the pipes don't align with the city streets,

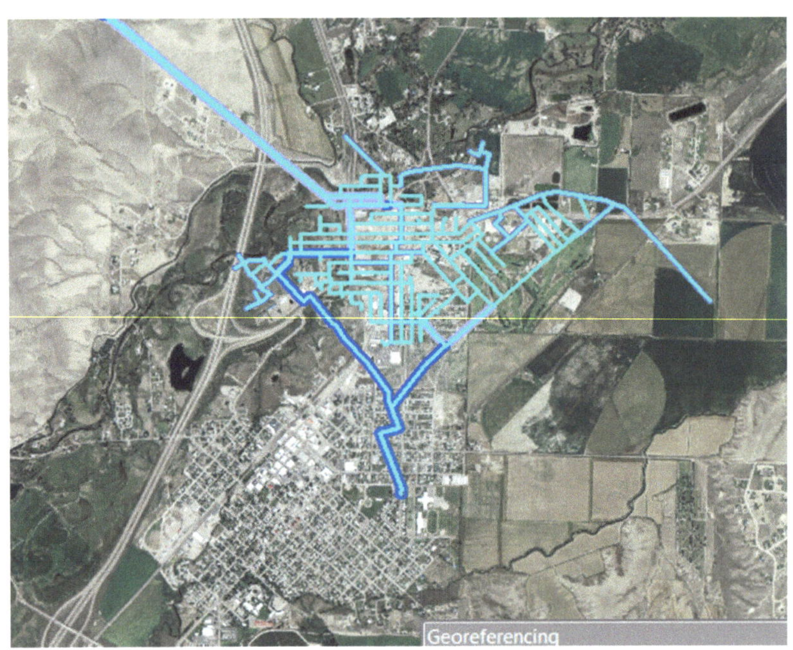

FIGURE 17 CAD GEOMETRY NOT GEO-REFERENCED IN GIS

then the CAD data must be spatially referenced to a coordinate system, datum, projection, and units that the GIS software can interpret. Ideally, the CAD data would have the same spatial reference as the other GIS data that they will be used with, but as long at the GIS software can interpret the spatial reference, and if the GIS software can project data on-the-fly, then the CAD data should align well with the GIS data.

In order to geo-spatially reference data, one must either collect all the data with geo-spatial coordinates, or tie a few control points to a known reference system then compute the spatial coordinates of the remaining CAD data. In this example we will perform a *geographic transformation* on the CAD in the GIS environment using the known state plane coordinates of some survey points. The transformation will translate, rotate, and scale the CAD data into geographic space.

ESRI's ArcMap provides tools for projecting CAD into geographic space, by entering coordinates for known control points, or by clicking on GIS features to associate CAD points to GIS points. For this project we have a CAD dataset which is in *assumed* (or arbitrary) coordinates, however, we do have some survey points in the CAD drawing that have state plane coordinates which we can tie to identifiable GIS features in the CAD dataset. We can use the survey points to geo-reference the CAD data by selecting each associated CAD point, then clicking on the survey point as a means to identify the *from* and *to* coordinates for a geographic transformation in scale, rotation, and transformation. Figure 18 shows the process for geo-referencing the CAD Dataset in ArcGIS.

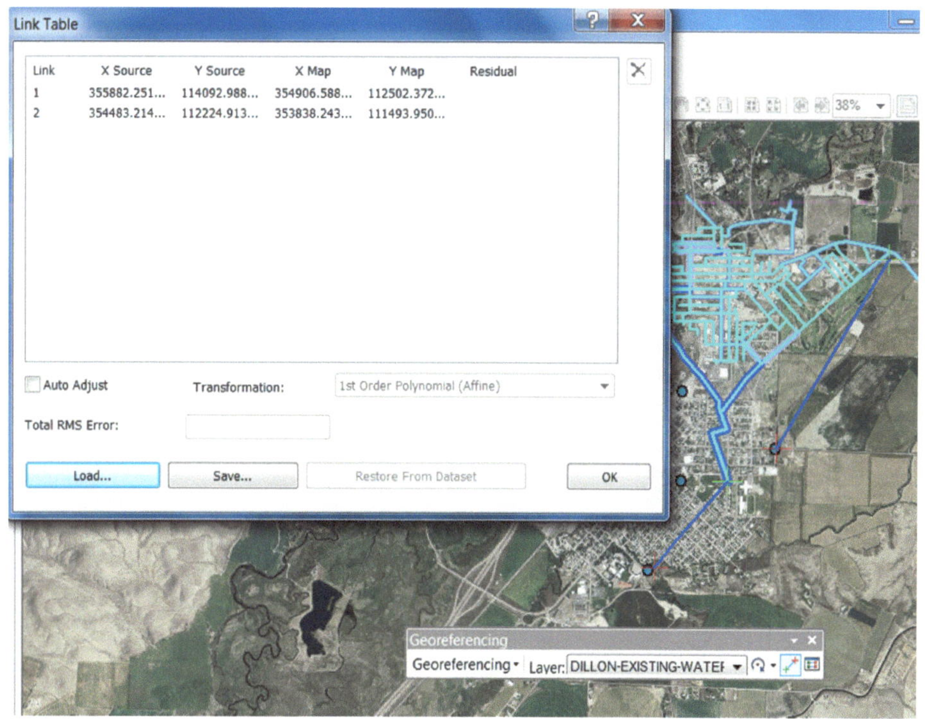

FIGURE 18 GEOREFERENCING CAD GEOMETRY

The

result of the geo-referencing operation is the CAD dataset placed geographic space, which the GIS software aligns with other GIS data. After geo-referencing the municipal CAD dataset into state plane coordinates, the water line features now align with the aerial imagery - see Figure 19 where the geo-referenced CAD data align (blue lines) run along the streets in the photography.

CREATING THE GIS DATA FROM THE CAD FEATURES

After the CAD file is geo-referenced, we can select the specific CAD entities that we wish to convert to GIS format, and then *export* those to a GIS format. In this project we select only those objects that were *polylines* contained in certain CAD layers because we are interested in only those CAD entities that represent pipes. We did this by querying the CAD database using a SQL (Structured Query Language) query, which queries the CAD database to identify and select all the features in the appropriate layers (Figure 20). We then export the *selected* CAD objects to a shapefile (GIS format).

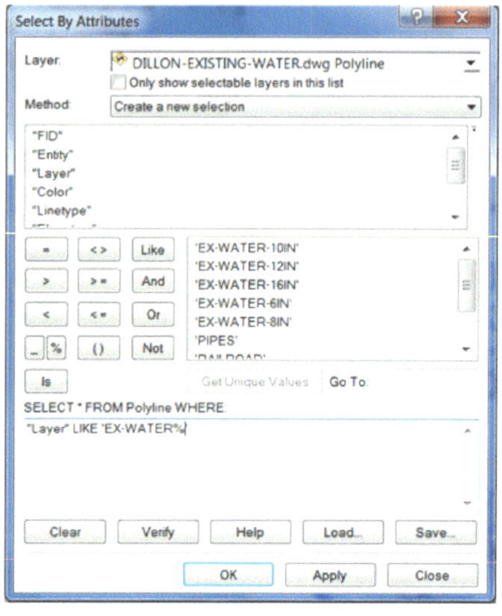

FIGURE 19 QUERYING THE CAD DATABASE TO SELECT CERTAIN ENTITIES

Notice in Figure 21 that the shapefile of the water lines contains some attributes based on the original CAD data. These are attributes that the GIS export operation automatically assigned to the dataset.

With a little bit more work, we could add other attributes which we could populate in order to facilitate symbolizin g, querying, analyzing

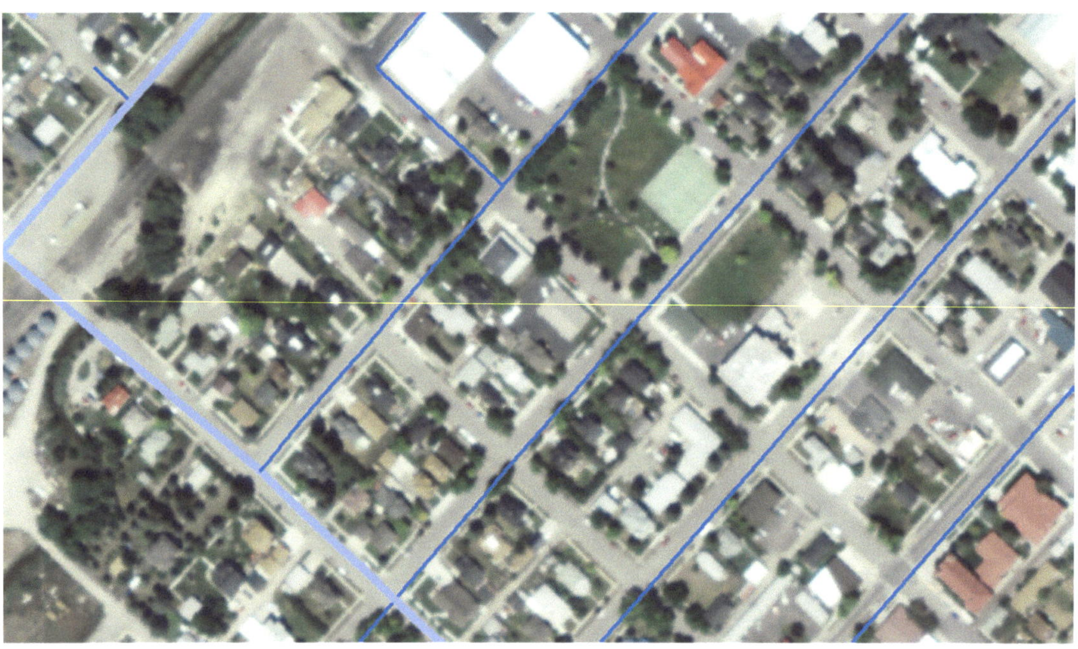

FIGURE 20 GEOREFERENCED CAD DATA (BLUE LINES) ALIGN WITH THE PHOTOGRAPHY

and displaying the data. For example, we might create a field called PipeSize to populate values based on the layer names that came from the CAD, or we could add a field for the date that the pipe was installed. Additionally, once we convert the CAD data to GIS format we can perform data integrity operations, such as running rule-based topological checks to ensure that the lines are connected, or we could perform SQL queries on the attributes to find invalid entries, etc.

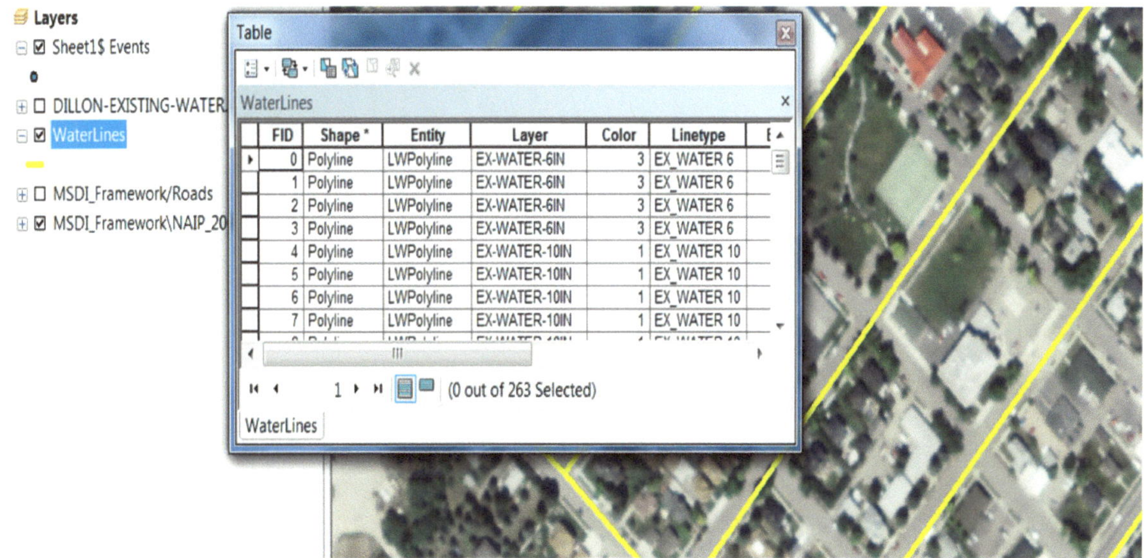

FIGURE 21 GIS FILE OF FEATURES (YELLOW LINES) DERIVED FROM CAD DATA, WITH ATTRIBUTES (SHOWN IN TABLE)

CHAPTER 10 DIGITIZING

Digitizing is one of the most important means to create GIS data from existing hard copy maps. Early GIS data conversion processes included taping a hard copy map to a digitizing board, then tracing the lines using a digital puck which collects coordinates. The tracing may have been done by clicking on endpoints of lines, which was useful for linear features such as boundary lines or by streaming coordinates as the puck was moved along line, which was useful for natural features such as rivers. However, most digitizing done today is performed on-screen from the scanned image of the hard copy documents (figure 22). This work goes faster and can take advantage of algorithms that automate the line tracing. Digitizing can also be used to create entirely new data based on visually identifying objects in existing GIS data or photographs.

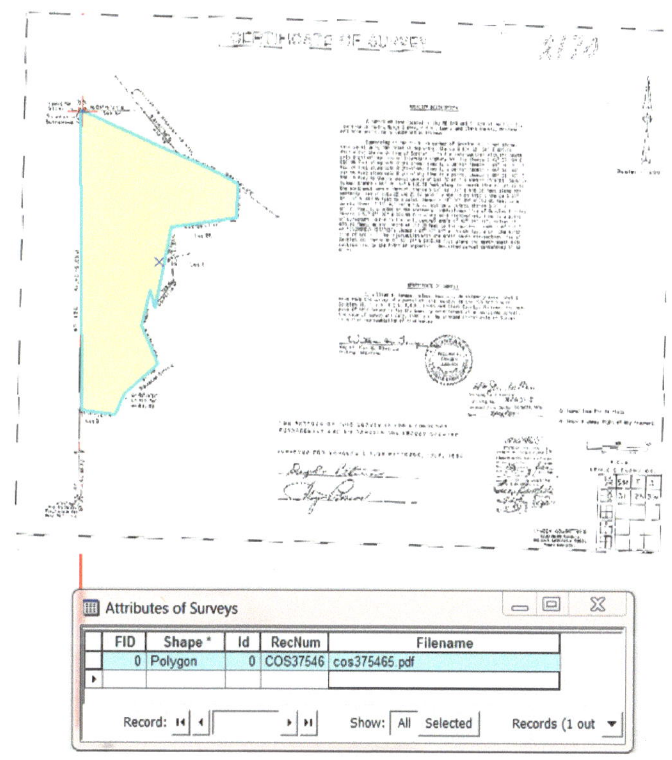

FIGURE 22 GIS DATA FROM SCANNED SURVEY

The key to digitizing is taking coordinate locations off a map to create GIS features. Therefore, in order to digitize an existing hard copy map, the scanned image must first be geo-referenced (translated, scaled, and rotated) into geographic space.

In many cases the new geometry may need to snap to the vertices of existing GIS features such as road centerlines, or parcel line vertices. It may also be necessary to close all polygons and ensure there are no loops, overshoots, or other topological errors in the polygons.

In some cases the GIS features may be built from existing GIS features of other datasets by copying their geometry and then pasting that into the new GIS dataset. This operation of editing by "Select-Copy-Paste-Merge" methodology can be a quick, efficient way to ensure that the new features align with existing features where appropriate. It is also possible to trace lines of existing features to create new features.

A more mathematically accurate means of digitizing existing maps can also be done using Coordinate Geometry (COGO). COGO is a method for creating lines and curves by typing in geometric measures, such as the bearings and distances of survey lines. COGO generally produces more accurate geometries than tracing scanned images as shown in figure 23. However, COGO takes more time and thus costs more. The needs of the project and the availability of the existing reference data, will dictate whether to trace, copy/paste, freehand sketch or COGO to create new GIS data from existing information.

These methods create the GIS geometry of features. The attributes (or tabular data) require other steps to get them into digital form and to associate them with the appropriate features.

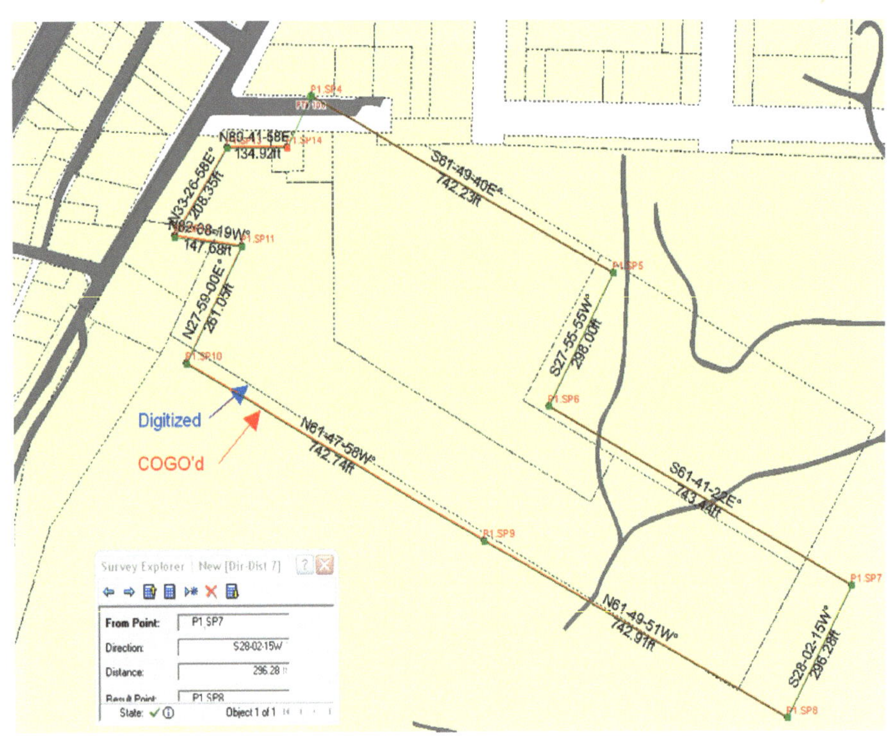

FIGURE 23 ACCURACY DIFFERENCE OF COGO VS. DIGITIZED

CHAPTER 11 GIS FEATURES FROM TEXT FILES

A lot of information is contained in simple text files, as common delimited or tab delimited data. It is possible to create GIS data from text files if the text file contains spatial information that can be plotted on a map. Typically text files will create GIS point features based on a single coordinate for each record in the text file, although it is possible to create GIS line or polygon features from ordered lists of coordinates.

If the text file contains other information in addition to the coordinates then it may be possible to create GIS attributes from them that are associated with each coordinate feature. The process described here works equally well with Excel spreadsheets with only minor differences depending on the GIS software one uses.

Text files are mostly stored in ASCII format. The ASCII (the American Standard Code for Information Interchange) format is a commonly used format for exchanging computer data. ASCII has been in use since the 1960's and has been used in GIS for about that long. ASCII formatted data files are a common export/import option on most surveying software as well. In the GIS realm ASCII formats are used for a variety of raster and vector data. GIS software may read some ASCII formatted data directly, or may import or export the ASCII formatted data to other GIS formats such.

Here we explain how to use an ASCII formatted *point* dataset in GIS and how to convert the ASCII point data to a GIS format (such as a shapefile or geodatabase feature class) with attributes using ESRI's ArcMap.

With ESRI's ArcMap one can easily load a delimited text file by using the tool Add XY Data. This will load the text file for display in geographic space provided that the text file has the following elements:

1. Each record is on a separate line.
2. The data for each record are separated by commas.
3. Coordinates for x coordinate and y coordinate are numeric values.
4. The coordinates may be geographic or projected coordinates.
5. Note that geographic coordinates cannot be in degrees, minutes, seconds (DMS) format - that format is non-numeric. If your coordinates are in DMS format, you must convert those to decimal degrees. Other numeric coordinate formats are OK so long as the GIS software

understands the spatial reference and units. Projected coordinates such as UTM meters, or State Plane feet will work as well.

6. The first row must contain the header information which means that the field names for each data element is listed here in the order that the data appear in each record. The header row identifies the field names including which fields contain the X (or easting) coordinate and Y (or northing) coordinate. The headers will be used as field names so they must conform to a ten-character field name limitation, with no spaces or special characters in the name.

7. The header row must be the *first* row in the file. If it is not, then delete the rows before it until it is.

8. *Optionally*, the file may have an ObjectID field as a numeric data type for unique identifiers for each record. If the ObjectID field exists, then you will be able to perform a few basic spatial functions on the data such as selecting points on the map. All other fields, as listed in the header row in the file, will be treated as attributes for which each record.

VIEW AND PREPARE THE FILE FORMAT

The sample file shown in Figure 24 is a comma delimited text file. That means that a special character, in this case a *comma*, defines the separation of data values between fields in each row. In addition, each row is treated as an individual record, so the number of row equals the number of records. If an individual record has no value for a particular field, then there will be no data between the delimiters. In this sample dataset that would look like two commas with nothing between them.

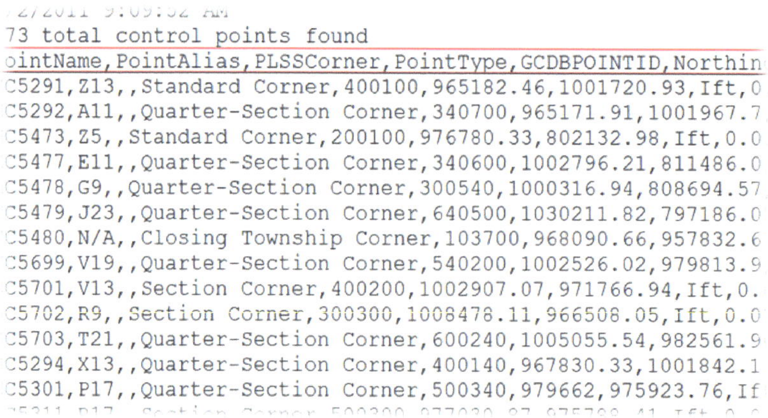

```
73 total control points found
ointName,PointAlias,PLSSCorner,PointType,GCDBPOINTID,Northin
C5291,Z13,,Standard Corner,400100,965182.46,1001720.93,Ift,0
C5292,A11,,Quarter-Section Corner,340700,965171.91,1001967.7
C5473,Z5,,Standard Corner,200100,976780.33,802132.98,Ift,0.0
C5477,E11,,Quarter-Section Corner,340600,1002796.21,811486.0
C5478,G9,,Quarter-Section Corner,300540,1000316.94,808694.57
C5479,J23,,Quarter-Section Corner,640500,1030211.82,797186.0
C5480,N/A,,Closing Township Corner,103700,968090.66,957832.6
C5699,V19,,Quarter-Section Corner,540200,1002526.02,979813.9
C5701,V13,,Section Corner,400200,1002907.07,971766.94,Ift,0.
C5702,R9,,Section Corner,300300,1008478.11,966508.05,Ift,0.0
C5703,T21,,Quarter-Section Corner,600240,1005055.54,982561.9
C5294,X13,,Quarter-Section Corner,400140,967830.33,1001842.1
C5301,P17,,Quarter-Section Corner,500340,979662,975923.76,If
```

FIGURE 24 EXAMPLE TEXT FILE OF POINT INFORMATION WITH HEADER ROW

Since our sample dataset has a couple extra lines before the header row we must remove the two extra lines at the top of the file. You can do this with any text editor such as Notepad, which can save in ASCII format. Be careful not to use word processing software that would change the file format.

ADD THE POINTS TO ARCMAP

After editing the file to clean up the data, if necessary, the file is ready to add to ArcMap. To do this in ArcMap we use **TOOLS --> Add X Y Data** (figure 25), then browse for the text file.

This utility allows one to *browse for the file, designate which field contains the X coordinates and which field contains the Y coordinates*, and specify the *coordinate system* of our X and Y coordinates. It is very important to specify the correct coordinate system, datum, projection, and coordinate units, because the spatial reference defines the proper placement of the points into geographic space.

After you set all the parameters in the Add XY Data dialog box and click OK, then ArcMap reads in each line of the file, and creates a point graphic on the screen for each record (if the coordinates make sense) as shown in Figure 26. The point graphics are based on the XY coordinates, and all other data are treated as attributes for each point.

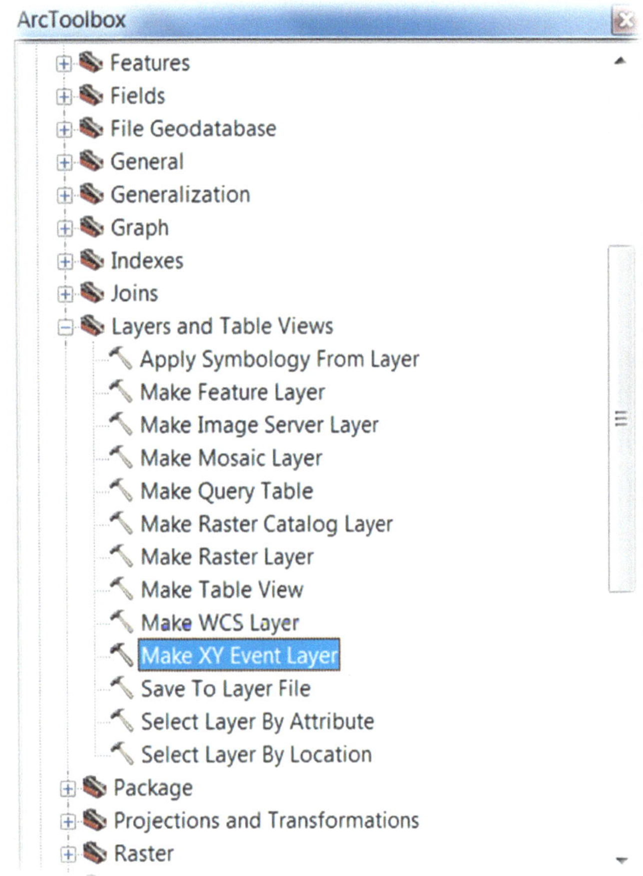

FIGURE 25 ADD XY DATA TOOL

The *Add XY Data* process adds the text file's point data as an *event* theme in ArcMap. Event themes are not a GIS type – they are a virtual geographical interpretation of the information in the text file which maintains a linkage to the original text file. The original text file is still intact with no changes made to it. ArcMap is merely reading and interpreting the text file for display in ArcMap. The graphics and the attributes are available for view only in the ArcMap project in which the event theme was created. It is not available as a GIS dataset for any other application. However, once you add the ASCII file to ArcMap as an event theme, you can add other GIS layers to the map and everything should properly align.

CONVERT TEXT FILE TO GIS FORMAT

To create a GIS dataset from the event theme, you must export the event theme to a shapefile or to a feature class in a geodatabase. To export to a GIS dataset, right click on the event theme layer in the map's table of contents, then select Export from the list of options. This will initiate the *Export* dialog (see Figure 26) which allows you to save the event them as a GIS dataset with the file name of your choice and, optionally, to change the spatial reference as well. The result of the export operation will be a new file such as a shapefile. This operation does not make any changes to the original ASCII file but instead creates a new GIS dataset such as that shown in figure 27. The new GIS file will be immediately available to use in any number of GIS display and analysis operations.

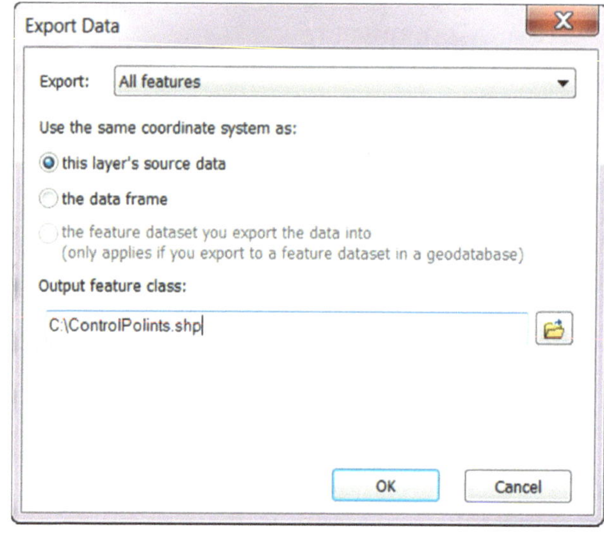

FIGURE 26 EXPORT EVENT THEME TO GIS FORMAT

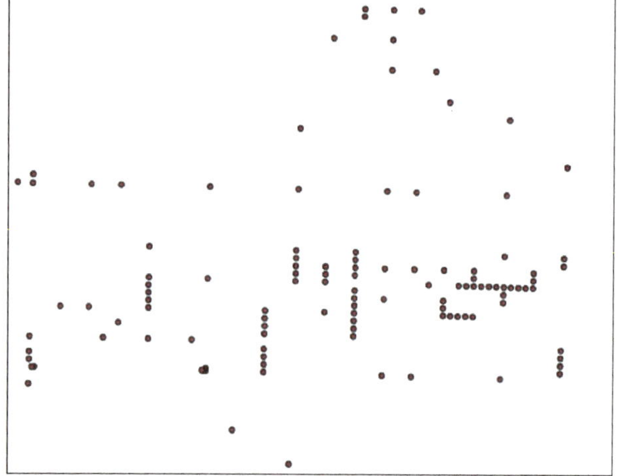

FIGURE 27 GRAPHIC POINTS AT EACH COORDINATE LOCATION

CHAPTER 12 GEOCODING – CREATING GIS DATA FROM ADDRESSES

One very powerful method of creating GIS data from existing information is address geocoding as in figure 28. Geocoding is the process of matching a property address against a reference file such as an address point file, a list of buildings that have coordinates or, more typically, a road network that has address ranges. The advantages to geocoding are that existing mailing lists, spreadsheets of addresses, databases that contain an address field, or any other source document, such as client lists, can be used to create a map layer.

FIGURE 28 GEOCODING TO CREATE GIS FEATURES

Once the map layer is created the power of GIS analysis and presentation becomes available. For example, a database or spreadsheet of company clients typically contains information about the client and the clients' location (i.e. address). If, for instance, a survey firm had an existing spreadsheet of the clients that it had over the past five years, and that spreadsheet showed the addresses where the work was performed, as well as the amount of profit, then creating a GIS layer from that spreadsheet would allow the firm to show where the jobs are coming from and what areas have been more profitable. That type of information is powerful when going after work. It could also help a firm decide whether or not to open a branch office in a particular area. Geocoding is used to provide driving directions in web mapping applications and car navigation GPS devices.

Additionally, geocoded addresses are often used to optimize routing. For instance, if a firm had a number of small jobs to perform in a day, it could geocode the location of the projects, and then performs a GIS route optimization program. In a matter of seconds, the GIS would report the order in which to do the jobs either to minimize the travel time, or to minimize the number of miles driven. The software will even provide turn-by-turn directions from one job to the next.

Geocoding requires a few simple things. First are the GIS software that can perform the address matching, and a reference layer, such as a road network that has address ranges. The second thing that is needed is a source file of addresses to map. Once a reference layer is set up, it can be used as many times as needed. So many different sources can be geocoded against a reference layer. Each source that is geocoded becomes a separate GIS layer.

In order to perform a geocoding operation one must first have a *reference* layer. The reference layer is an existing GIS layer that contains address lookup information such as road names and address number ranges. The most commonly used reference layer is a road centerline. Road centerlines will contain addressing information such as street names, address ranges for each side of the road, postal codes, state or province name, and city name. Road centerlines can be obtained from a few sources. Generally local governments (cities or counties) create and maintain them, the US Census Bureau TIGER line files can be used, and there are many commercial products, such as well which may have more enriched data associated with address.

In GIS one must first set up the reference layer (may also be called the Address Locator) in order tell the software which file you want to for the reference layer, and how that particular file stores the address information (i.e. what the fields are called for the various address components. Once the reference layer is set up, it may be re-used whenever needed.

Facility	Address	City	ST	Zip
Fallon Medical Complex/Hosp	PO Box 820	Baker	MT	59313-0820
McCone County Health Ctr	PO Box 47	Circle	MT	59215
Roosevelt Memorial Med Ctr.	PO Box 419	Culbertson	MT	59218-0419
Dahl Memorial Healthcare	PO Box 46	Ekalaka	MT	59326
Rosebud Health Care Center	PO Box 268	Forsyth	MT	59327-0268
Frances Mahon Deac. Hosp	621 3rd Street S	Glasgow	MT	59230-2604
Glendive Medical Center	202 Prospect Drive	Glendive	MT	59330-1943
Garfield Co. Health Center	PO Box 389	Jordan	MT	59337
Holy Rosary Health Ctr.	2600 Wilson	Miles City	MT	59301-5094
Phillips County Med Center	PO Box 640	Malta	MT	59838-0640
Sheridan Mem. Hosp. & NH	440 W. Laurel Avenue	Plentywood	MT	59254-1526
Poplar Community Hospital/NorthEast MT	315 Knapp Street, PO Box 38	Poplar	MT	59255-0038
Daniels Memorial Hospital	PO Box 400	Scobey	MT	59263-0400
Sidney Health Center	216 14th Avenue SW	Sidney	MT	59270-1690

After a reference address layer is set up then one can use it to create GIS data from lists of addresses. Figure 29 shows an example address source file which, in this case, is a spreadsheet of names of health care facilities and addresses.

The address information is contained in multiple fields in this table which is in a manner often found in legacy datasets. The address fields in this example are called Address, City, ST, and Zip. Depending on the reference file's address information the fields in this source file may or may not work. Note however, that some addresses are PO Boxes. These cannot be geocoded because they are not physical addresses.

Correct formatting for geocoding depends upon the format of the reference data because the geocoding process attempts to find matches between the source data and the reference data. For example a house number and road name in the source file, must fall within a valid range of numbers along a segment of road in the reference data. For instance, if the address is 150 W Main Street, then it would fall halfway along the 100 block of W. Main St. The side of the street that it would be on depends on the local convention for even and odd ranges along a road, provided that information is encoded in the reference file.

FIGURE 30 GEOCODED POINTS

Once the source file is set up, the geocoding operation is run. The operator may need to adjust the sensitivity for spelling and what is considered a 'match', i.e. the level of guessing that the software is allowed to do. To deal with the failed records may require direct user matching, adjustments in the sensitivity, corrections on the source record, or accepting the rejects.

The product of the geocoding operation is a GIS point file that contains all the field (attribute) information of the original record. Figure 30 shows the addresses plotted on a map (green dots) over an aerial photograph. Each dot represents a record from the source file. Each point feature has associated attributes from the original table. Because the data from the table are now attributes of the GIS features, analysis can be performed on the GIS point file. Additionally, the features can be symbolized based on attribute values.

As mentioned earlier, the geocoding procedure is very simple and quite powerful. Any properly formatted address information can be geocoded against a standardized reference theme. Where you take from there depends on the data contained in the source table and your own imagination.

CHAPTER 13 GNSS RESOURCE MAPPING FOR GIS

Global Navigation Satellite Systems (GNSS) such as the Global Positioning Satellite System (GPS) of the United States of America (figure 31), provide positioning resources that can help one to rapidly map objects on the ground. Resource grade GPS is the term used for mapping features to accuracies outside the typical surveyor boundary location and control network accuracies. Survey grade GPS and terrestrial survey equipment can yield measurements accurate to centimeters or even millimeters. Resource grade GPS equipment used for feature mapping is accurate to within meters or in some circumstances, decimeters.

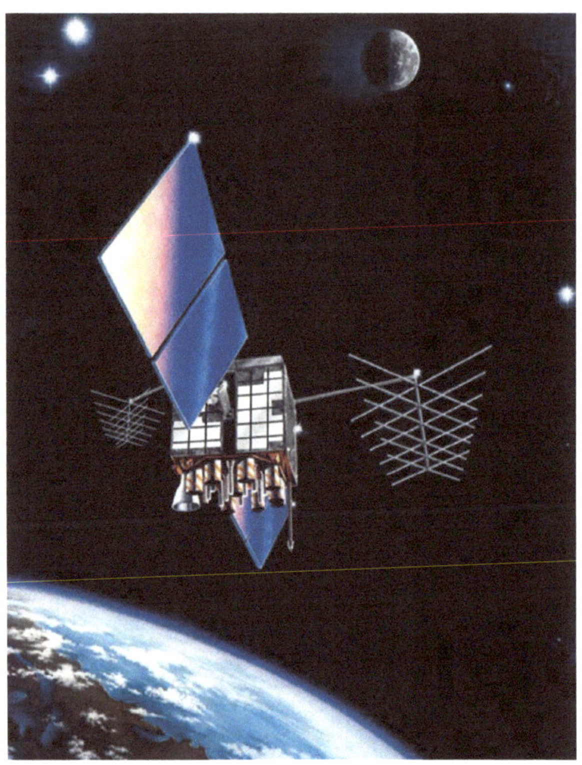

FIGURE 31 GPS SATELLITE

There is a lot of resource mapping being performed for GIS. The GIS community is mapping fire hydrants, soils, wetlands, water lines, manholes, historic sidewalks, water tanks, and many other features. Additionally, utility companies are mapping the transmission and pipelines as well as other important facilities. One of the most important resource mapping efforts is road mapping done in support of routing, addressing, and emergency response. Thousands of miles of roads, driveways, and structures have been mapped to support rural addressing so that the Enhanced 911 (E911) could be implemented. E911 provides the address of the telephone number on the 911 dispatch center's screen when a call comes in.

Surveyors bring a unique perspective to resource mapping because surveyors are trained and experienced at collecting field data and assessing the spatial accuracy of those data. Surveyors know how to collect data under nearly every kind of environmental, topographic, and other field condition, and can perform this work expertly and efficiently. It is common for some GIS folks to perform mapping without having a process built for uniform collection of all the data they need to collect. The results can be: 1. Not all the features were mapped, resulting in incomplete data; 2. The data collected were described inconsistently, resulting in many different descriptions for the same thing; 3. Confusing and disordered data, resulting in descriptions for one thing appearing in the wrong place, such as putting tree species in the tree diameter field, or calling a road surface such things as concrete, arterial, public, and highway. Work that is not performed consistently becomes useless, and correction of the data may require re-mapping the data in the field.

IMPORTANCE OF ERROR CHECKING AND QUALITY CONTROL

Surveyors are also experts in error checking for spatial accuracy, data completeness, and attribute accuracy. Quality control procedures are routinely performed by surveyors. Because of the importance of error checking and quality control, these procedures are part of every surveyor's training, both in the classroom and on the job. Unfortunately, this component of mapping is not always included in GIS projects. GIS folks come to GIS from a variety of backgrounds, such as computer science, natural resources, and planning, and may not have had the quality control and error checking training that is important for mapping projects.

On the other hand, the surveyor's background could also prove to be an impediment to successful completion of GIS projects unless the surveyor understands GIS well enough to meet the client's needs. One of the most important things that the surveyor needs to understand is the client's accuracy requirement for the mapped data. The surveyor should work with the client so that they both understand how the data are to be used, and how it will be integrated with other information. With that understanding, the long term needs, and the immediate considerations of cost and the project time frame can be met. Sometimes this process is an education for the client; sometimes it is an education for the surveyor. Usually both parties come away from the experience with increased knowledge about the process.

The general steps to performing a GPS resource mapping project are listed below

1. Project Plan
2. Perform Field Work
3. Process the Field Data
    - 3.1. QA/QC for Completeness, and Attribute Information
    - 3.2. Differential Correction (if not RTK)
    - 3.3. Error Checking
    - 3.4. Export to GIS
4. Build Metadata

STEP 1: PROJECT PLAN SCOPE

The first step in any project is, of course, planning. It is important to understand the scope of the project, what data to collect, and the approximate size of the project. Another consideration is which data to collect in the field, and which data to enter in the office. Many projects already have a set of data in a spreadsheet or database or some digital format, but they are lacking a map component. For instance, a Water Quality Protection District had a public water supplies data base with a lot of statistical data on the water supplies (fluorine, chlorine, sulfur, nitrates, etc.). However, since there was no geographic component, there was no way to map the data for analysis and study. In that case, all that was needed was to GPS the approximate location of the water supply and record its unique identifier from the database. Once the GPS work was complete, the database information was joined to the GIS features, and any of the statistical data could be mapped and analyzed. One of the map products from this project was a map of fluoride levels of the area, which was generated from the point source data (basically this is the same process as generating contours from random elevation points).

In other cases the information in an existing database may be out of date, in which case a field inventory is necessary to update the data (or to collect missing information), in addition to mapping the location information.

Questions to ask prior to undertaking a resource field inventory are:

- Is there an existing digital dataset for these features that could be used?
- Is there existing attribute data that can be joined to the mapped feature?
- Should all the information be collected in the field or only the coordinates and ID?
- Should some

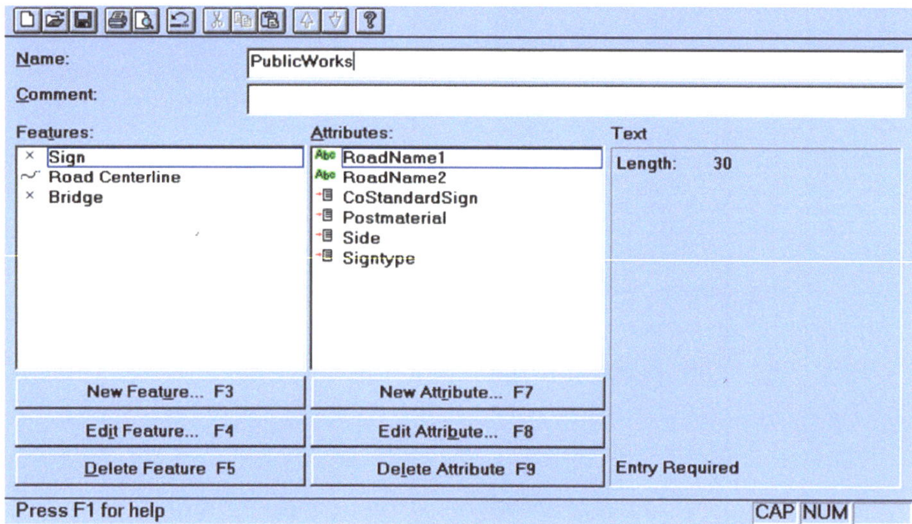

FIGURE 32 EXAMPLE DATA DICTIONARY – MAPPING TEMPLATE

new data entered in the office?

- Will the data be maintained, or will this be a static dataset i.e. a snapshot in time?
- How will the data be used? Is it just to make a single map, or is it for analysis as well?
- What is the long term objective for the data?

Data Dictionary

The data dictionary is the template for data collection (figure 32). One should use a data dictionary to make the data collection simple and consistent. The data dictionary defines what will be mapped, how it will be mapped, and the characteristics (attributes) to be collected for each feature. For example, a data dictionary for an address mapping program might include the road name, road beginning point, road end point, address ranges for each side of the road, the city name, the location of road intersections etc. Most data collectors use pick lists for features and attributes which can be set up in the data dictionary. Pick lists ensure consistent descriptions for features and attributes which is important for GIS data representation and analysis. An example pick list would be tree species for a tree inventory, and perhaps valid ranges for tree diameters.

GPS Mission Planning

Although the current satellite constellation provides nearly round-the-clock mapping opportunities, it is still important to check the availability and configuration of satellites in the project area for the time one intends to perform the field work as shown in figure 33.

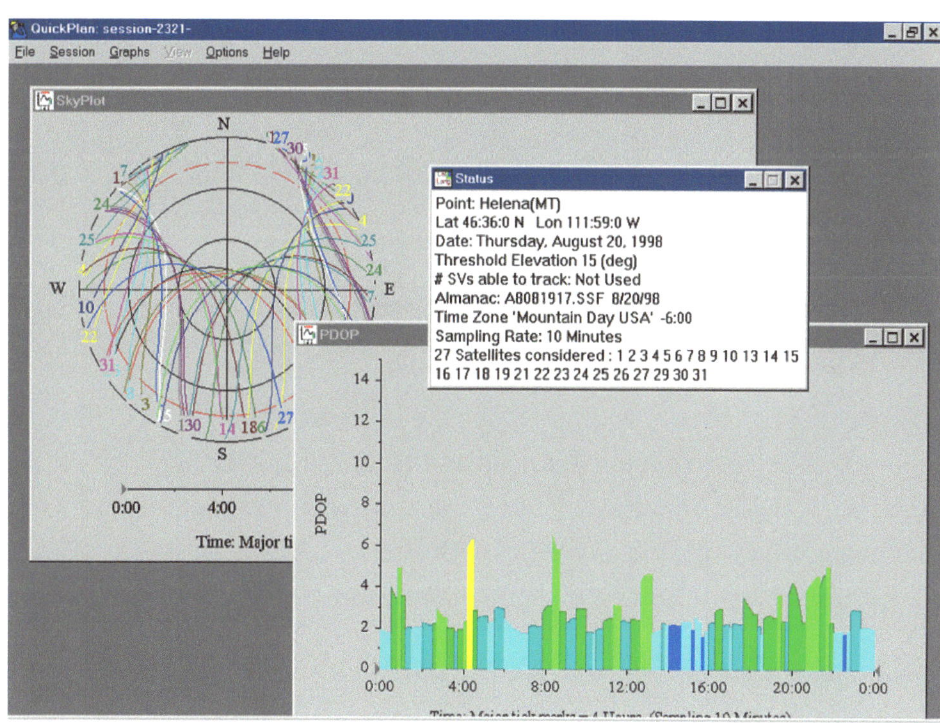

FIGURE 33 GIS MISSION PLANNING

Most GPS software provides a mission planning module which uses the latest satellite almanac to estimate satellite configurations for the project location, date, and time. Use this information to preplan lunch breaks and travel or setup time during times of high PDOP (position dilution of precision—the higher the PDOP, the worse the conditions). Mission planning is particularly critical in canyons (including urban canyons) and forested areas where the GPS receiver may not be able to "see" all available satellites.

STEP 2: FIELD WORK

Field work for GIS is not any different than other types of surveying field work; therefore surveyors need no instruction here. Check in on a few known control points if possible so that the positional errors can be tested.

STEP 3: PROCESS THE FIELD DATA

Data Download

Data downloading is the process of moving the data from the field data collection device to the computer for further processing. Usually this is a simple file transfer process once the devices are connected and communication protocols have been established. It is good practice downloaded daily and back up the raw data files. This protects a day's work and clears out the data collector's memory for the next day's work.

QA/QC for Completeness, and Attribute Information

Once the data are downloaded and backed up, a quick check of the data should be performed while the day's work is still fresh in the operator's mind. Check to ensure that there are data for all the features that you believe you have collected and fix any data entry errors for attributes or features.

Differential Correction (if not RTK)

Differential correction, which reduces the location errors, should be done either in the field or in the office, depending on equipment used and resource availability. This operation has become quite simple and quick. This is basically a data comparison between your new data on unknown points, and data collected at the same time on a point with known coordinate values. This will put your point data closer to the "actual" location and improve the geometry of line and area features.

Error Checking

Quality assurance and quality control for spatial errors is often neglected in GIS. Regardless of the spatial accuracy required for the project it is important to know what the magnitude of the error is. Accuracy estimates are standard metadata elements. Also, sometimes the spatial error will be larger than expected, in which case the data might not be useful. The magnitude of the error becomes part of the metadata for the project.

Export to GIS

After the data have been checked for completeness, consistency, accuracy, and the magnitude of the spatial error determined, and then the data can be exported to the target GIS format (figure 34). The two important considerations here are the GIS file format, which depends on the GIS software to be used, and spatial reference for the GIS dataset. Most GPS software allows exporting the data to a different spatial reference than that in which it was mapped. The spatial reference must be reported in the metadata report.

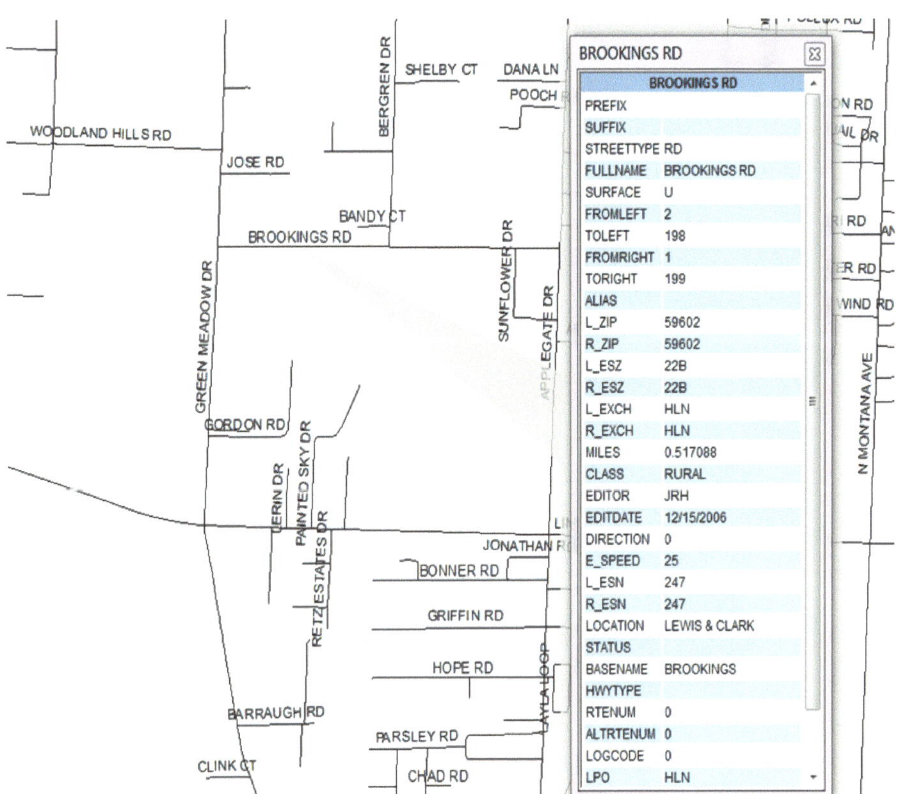

FIGURE 34 MAP SHOWING GIS ROADS DATA CREATED BY GPS

STEP 4 METADATA

This step is often the last, and all too often neglected, but metadata are an essential the report about the data. The metadata provides the users of the data with important information about the data, when, why and how it was collected, and by whom, which scales the data are intended to be used at and other important aspects of the data so that those who wish to use it may determine its usability for their purpose. Many of these items that go into the metadata are decided during the project planning stage, so that is a good time to start working on the metadata. In fact, a metadata template for the project can be used as a guide for the mission planning.

CHAPTER 14 DIGITAL ELEVATION MODELS

A digital elevation model (DEM) is a numerical representation of terrain elevations. The terrain is sampled at intervals and the sampling may be done by field survey methods or remote sensing such as Lidar or photogrammetry, or digitizing elevations off maps.

FIGURE 35 HILLSHADE FROM DEM

The elevation data may be stored in a text file or grid format of coordinates with corresponding elevation values. The coordinates can be listed in a variety of geographic projections and/or coordinate systems, and the elevation data can be displayed in any numeric unit. GIS software converts these elevation data into graphical representations for a visual display that helps one to "see" the lay of the land. A typical representation is a hillshade as shown in figure 35, which simulates heights on a continuous surface by calculating light and dark regions based on a sun height and angle. Additionally, DEM data supply important information for GIS analysis and statistical calculations.

HOW DEMS ARE USED

DEMs provide elevation data for such things as stream flow calculations. Fire risk modeling takes advantage of slope (steepness) and aspect (north-, south-, east-, and west-facing) analysis derived from a DEM. DEMs combined with surface and sub-surface hydrologic data are used for substance transport calculations for environmental hazard analysis. They are also used for orthorectification of digital aerial photography, subdivision planning, road building, overland right-of-way planning, and cell tower placement.

DEMs can be used along with radio power data to analyze areas of coverage by radio towers, and to determine the optimum location for new towers. This allows agencies to maximize communications in mountainous areas, and to minimize the radio "holes" where signals cannot be received.

There are many practical applications for DEMs. One very interesting application is to use a DEM to create a shaded relief map base for other digital data. A shaded relief map helps the viewer to really see how roads, rivers, property boundaries, and other map features are situated upon the topography. Looking at a flat map just doesn't help the mind to "connect the dots" quite like the same map overlaid on a shaded relief. With a shaded relief map such as shown in figure 36, the user can see that a bend in the road goes around a hill, and that one road goes up a gulch, while another one follows a ridge.

FIGURE 36 SHADED RELIEF MAP WITH SURVEY GRID (PLSS) OVERLAY

DEMs can also help when designing a survey control network. By overlaying survey control onto a DEM one can readily see the terrain where the control points fall.

One may pick spot elevations, create contours, create profiles (elevations along a line), calculate slope (percent or degrees), create a hill shade for terrain visualization, or perform aspect analysis (calculate which direction a slope faces).

OBTAINING AND INCORPORATING DEM DATA INTO GIS

Digital Elevation Model data are available for download free for many parts of the world. In the US the U.S. Geological Survey (USGS) offers digital elevation data at a variety of scales for most of the continental U.S., Alaska, and Hawaii as well as global elevation datasets http://eros.usgs.gov/#/Find_Data/Products_and_Data_Available/Elevation_Products . In other areas of the world data may available on a country-by-country basis though not necessarily free for download. In addition to data for downloading, there are some resources for connecting to map services which can display digital elevation data in GIS map viewers.

All DEM data is similar in logical data structure and is ordered from south to north in profiles that are ordered from west to east. The accuracy of the data varies, but the finest data has a vertical accuracy of about 15 meters, so these products are useful for projects at 1:24,000 scale or smaller.

CHAPTER 15 GIS DATA FROM LASER SCANNING & LIDAR

Scanning, terrestrial and airborne, has become an important tool for surveying in applications where the rapid collection of a high density of measurements is important, and is especially helpful in areas where physical access is difficult or dangerous. The products one can derive from laser scanning and Lidar are important in GIS.

FIGURE 37 LASER GENERATED DIGITAL SURFACE MODEL (DSM)

Lidar technology generates three dimensional point clouds of x, y, z points by measuring the time to return a laser pulse from the scanning instrument. The individual measurements are translated into geographic coordinate space to produce a dense set of measurements that can define objects in the environment based on their ability to reflect the laser pulse. Laser scanners can make millions of measurements very quickly which can then be used to create a variety of products such as the digital surface model shown in figure 37.

Measurements can be accurate to within millimeters or centimeters and hundreds of measurements can be made per minute. The data are collected as three dimensional point clouds of Cartesian (X, Y, Z) or polar (r, Θ, φ – range, azimuth angle, elevation angle) coordinates along with a measure of the strength of the return signal as an intensity value. Some instruments collect multiple returns of the same signal, which can help to further differentiate types of objects (such as water surface and subsurface measurements). Most laser scanners now include an integrated color camera in order to also collect pixel color and intensity values.

A laser scanner consists of a laser pulse generator that sends a laser beam out. The beam is reflected off the target surface and returns to the scanners where it is sensed by the beam receiver. The time that it takes to return to the scanner is used to calculate the distance to the target object. Thus, laser scanner measure without physical contact. Their range is up to 1 kilometer. The collected data can be very large files, but the measurement density can be controlled to adjust the accuracy and file sizes.

DATA FORMATS

There exists a Lidar data *interchange* format standard - the LAS (LASer) file format for a three-dimensional point cloud. Specific software vendors may introduce other formats specific to how they manage and represent these large data sets. GIS software can help one to manage, store, retrieve, filter, visualize, analyze, and derive additional products from the basic point cloud data. GIS is very helpful in integrating point clouds with other geographic information in order to provide context and additional understanding of the point cloud data.

Types of applications for laser scanner include mapping dams, bridges, highway interchanges, busy intersections, archeological sites. Laser scanning helps for the following situations:

- when a lot of detail is required
- high accuracy is required
- complex geometries such as industrial plants with many pipes, cables and other objects
- Inaccessible areas such as busy intersections, dams, power lines, cliff faces.
- Safety issues are a concern

FILTERING

Point filtering will display a subset of points based on some predefined criteria. This helps with visualizing certain types of objects such as buildings or vegetation and can speed data processing by reducing the number of points to work with and speed up processing. A common filter is ground point filter in order to model the bare earth terrain. Ground point filtering uses only the Lidar points flagged in the point cloud file as *ground points* to display the LAS dataset. By contrast, non-ground filters use all the Lidar points that are not flagged as ground points to display the LAS dataset.

Since the data may contain multiple returns from a single pulse it is possible to track which the return number which can then be used to extract additional information about the measured objects. This can help, for instance in identifying vegetation (as first returns) in forested areas that are mapped by airborne Lidar.

CLASSIFYING

The data can be classified based on known common parameters such as return number, number of returns, intensity levels, etc. possibly supplemented with operator inspection of the data location and shapes. If the dataset contains classifications, then those classes can be displayed to show for example buildings, versus, ground, vegetation etc.

TYPICAL PRODUCTS

The point cloud measurements can be analyzed and processed to generate derivative products. Typical products are *Intensity Return Images*, *Digital Surface Models*, *topographic maps*, and Digital *Elevation Models* to support orthorectifying aerial photography.

Intensity Return Images are derived from intensity values returned by each laser pulse. The intensity values can be displayed as a gray scale image. These help to identify objects based on their reflectance as some objects reflect the laser light better than others. For instance, asphalt is a poor reflector, so paved roads can be identified based on the range reflectance values.

Topographic Lidar systems produce surface elevation as x, y, z coordinate data points. There are many topographical products that can be derived from raw point data. Common examples which can be generated and used in GIS software are

- Digital Elevation Models (DEMs)
- Digital Terrain Models (DTMs) -bald-earth elevation data
- Digital Surface Models (DMS) – bare earth and built environment (figure 37)
- Triangulated Irregular Networks (TINs)
- Breaklines - a line representing a feature that you wish to preserve in a TIN (example: stream or ridge)
- Contours
- Profiles (figure 38)
- Slope & Aspect

In addition to the point cloud data, some Lidar providers collect digital color or black-and-white

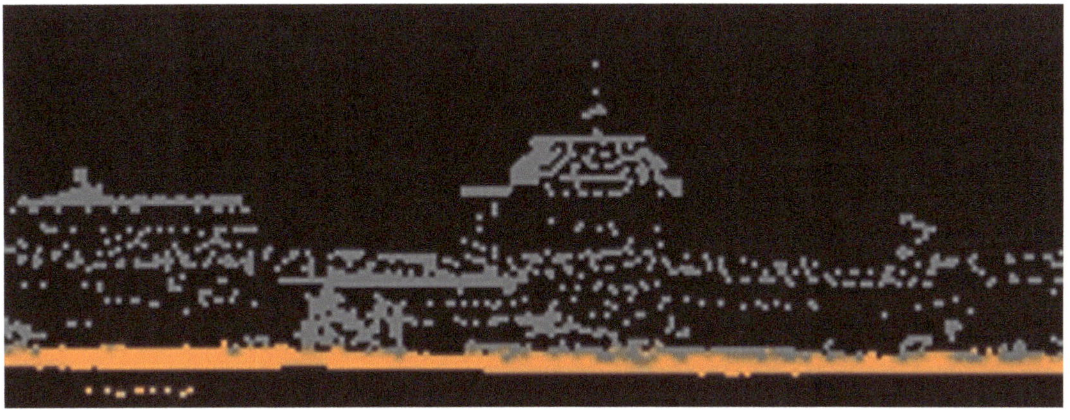

FIGURE 38 EXAMPLE POINT CLOUD PROFILE

orthorectified imagery simultaneously with the collection of point data. Imagery is collected either from digital cameras or digital video cameras and can be mosaiced. Resolution is typically 1m.

PART 4 GIS IN THE SURVEY OFFICE

This part introduces some ways that GIS helps surveying activities that occur in the office. We begin with an introduction to data modeling concepts as they apply to information and data sets relevant to surveying. We then examine a few specific data sets that surveyors use, then we how GIS helps with some common surveying endeavors.

CHAPTER 16 MODELING SURVEY INFORMATION FOR GIS

In order to store survey information electronically, the data must be modeled in some way. A data model such as shown in figure 39, is a logical structure for the storage of information, and determines what information is stored, how the data are represented, and how each piece of information relates to other pieces of information within the dataset. The data model may also determine how this dataset relates to other datasets.

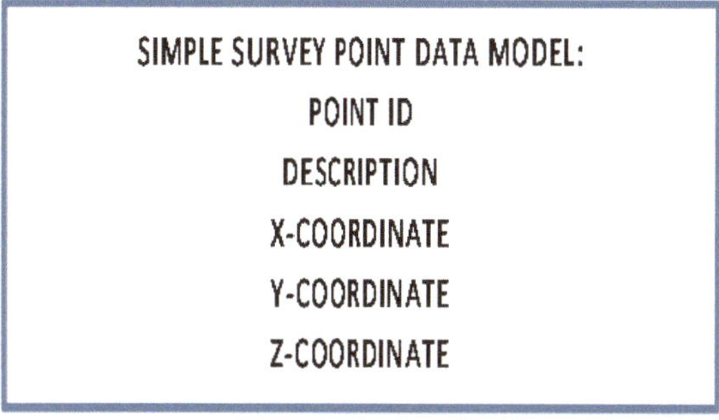

FIGURE 39 SIMPLE SURVEY POINT DATA MODEL

The concept of a data model should not be confused with the data format. The data format is the electronic storage form. Data formats include such electronic forms as ASCII, Dbase IV, XLS, etc. The data model, on the other hand, is how the pieces of information are structured, regardless of the data format. Some database management systems constraint the content and form of data in very structured ways, while others allow a variety of structures. A good data model is storage-format independent, and can be ported to a variety of formats. A very simple example data model for a survey point might be a flat structure that has no related tables or relationships to any other data, and which contains only a few simple pieces of information per record. Figure 39 shows such an example point data model.

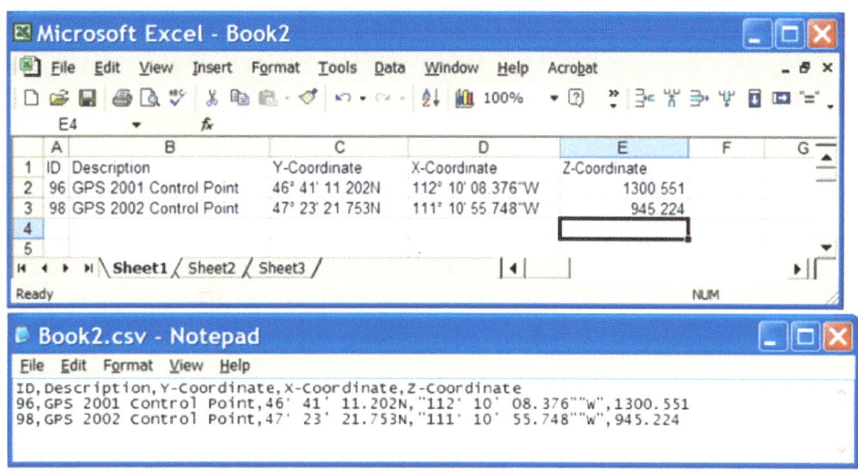

FIGURE 40 TWO FORMATS – EXCEL & CSV

At its simplest, survey data may be stored in a straight string of information. Typically, a text file of survey data uses one line of text for each record, where a record represents, for instance, a measurement, or characteristics of a survey point. Such a data model can be formatted as a text file, a spreadsheet or any number of other formats. Figure 40 show this simple data model in two such formats ASCII and Excel.

The difference between a data model and data format can be explained using a grade stake as an example. In grade staking there are certain kinds of information to convey and a medium used to convey that information. The wooden stake is the medium or format. The information that is written on the stakes is modeled in a particular way. The client will typically require a particular format for the data to put on the stake, how many stakes to use, where to locate the stakes, etc. We may choose to model our construction information in a two-stake format using a grade-stake and an offset-stake (shown in figure 41). Each stake contains its own information, and has a relationship to the other stake. The data model dictates what information to put on the stake and how that information is written (represented) on the stake. So long as everyone understands the data model, then the communication will be effective, which in this case is to communicate how much material to remove from a particular location.

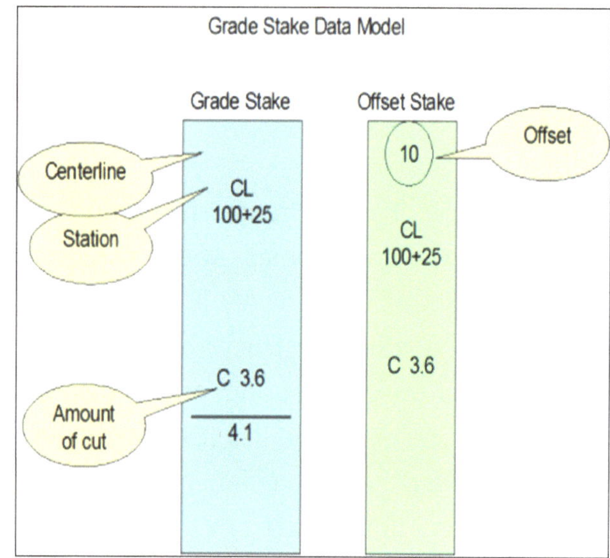

FIGURE 41 GRADE STAKE EXAMPLE

Thus, in general, the data model is about which information that we wish to convey and how we chose to convey the information. Data models can be simple or complex. The complexity of the data model depends on the nature of the information that is modeled and the intended use or uses for the information. In the example of the survey point in figure 39, the uses for the information may be limited due to the small amount of information contained in the data model. The use of the data might be to answer such simple questions as how many survey points do I have and where are they? But as we shall see later on, survey data can be modeled in very complex ways as well. The key to good design for data modeling is to understand the data itself and the applications that the data will support (i.e. uses of the data).

A data model may include a set of related tables with different kinds of data in each table. The tables may be related through the use of keys (unique identifier codes) that are common in different tables. A multiple table format in a relational database structure can reduce redundancy in the database and reduce the volume of data in the database. For example, a database might have a single list of survey points, and a single list of projects. The survey point, however, might be used in many projects. Instead of repeating all the information for each survey point within each project table, there may be only one set of control point records which are related to the projects through each survey point's unique ID, or key. Thus only the key needs to be listed in the project – not the full control point record. The application can then go to the survey point table to look up the other information about the point.

As we discuss GIS applications for surveyors in this section we will describe additional examples of how to model surveying data for GIS.

CHAPTER 17 CREATING A GIS DATABASE OF SURVEYS

Whether you work in a county surveyor's office or a private survey firm, you can create a GIS layer of surveys (figure 42) that can be used for a variety of purposes. A GIS survey layer can help one search for a particular survey to see where it is located. A GIS survey layer overlain on a parcel ownership GIS layer can help one find who the property owners are near the survey – perhaps to produce a contact list. One may be able to readily see the extents and chronology of surveys in an area or use the survey layer to find control points within a certain distance. Having a survey layer already mapped in place can speed up the process of mapping parcels that are built from surveys, or in cases where parcels have already been mapped the surveys GIS can help one find errors in the parcel mapping. A GIS layer of surveys can also provide a graphical index to a database of survey information and even the scanned image of the survey document such that access to that information is through a simple click on the survey GIS feature.

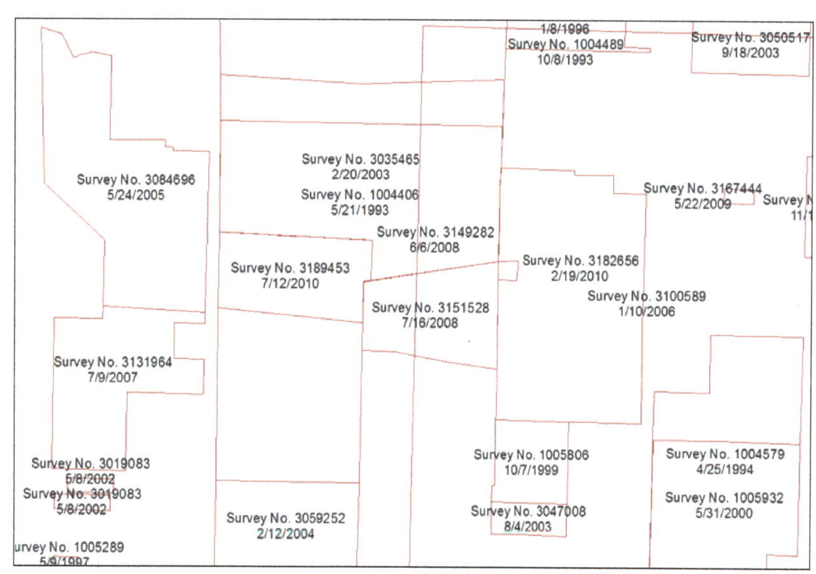

FIGURE 42 GIS LAYER OF SURVEYS

There are many different ways to create a survey index layers. The ideal scenario might be to COGO in the perimeters along with some GPS ties for of the corner positions of the surveys. While COGO is the most accurate method, it is usually impractical due to the expense and time required for the field work and COGO. Also, the payoff may not be high enough to justify the effort. That may depend on the area that the surveys are. In urban areas COGO combined with field ties may be necessary, because of the density of surveys, and the small size of parcels, streets, and other features in close proximity. Rural areas, on the other hand, tend to have large parcels and fewer surveys, and so they may not require such painstaking effort. Other quicker methods, such as described here, may be in order and still may yield a satisfactory product. Next we discuss a quick and very useful method that may not be viable for all situations, but can certainly achieve a high rate of production, while yielding good results. It is important to remember that regardless of the data conversion method one choses to create a survey GIS layer, the GIS will still provide a means to access the source documents and database. So if one needs more than a general view of the location and approximate extents of a survey, that information is merely a click away.

Generally the requirements for a GIS layer of surveys may be:

- ✓ The GIS layer will be a vector polygon layer to allow for spatial query and analysis.
- ✓ The surveys will be created as polygons (even if the survey itself is linear).
- ✓ Only the bounding perimeter of the surveys will be digitized – not any of the internal lines.
- ✓ Each survey feature (polygon) will have at least two attributes: record number (unique ID which may be based on the public record index number or office filing number), and a file name for the source CAD drawing or scanned image of the survey (if one exists).
- ✓ The source documents (either CAD drawing or scanned image) will be on-screen digitized in a coordinate reference system (such as UTM, State Plane, or local datum).

Some additional considerations are that all the survey perimeters will be digitized, even where surveys overlap. Other survey features on the survey map, such as corner monuments and control points in the survey will not be captured here. Those points might be captured in a separate GIS layer(s) but are not relevant to this discussion. Here we will use ESRI's ArcGIS software to create the editing environment and to perform the data conversion.

The first step is to set up an ArcMap project with the reference data which helps us to locate the surveys in geographic space, and helps to guide the placement of the survey boundary geometry (figure 43). In this case we have a dataset in the Western United States so a layer of Public Lands Survey System information (township, range and section and PLSS survey points); along with an existing parcel layer will serve as the reference layers. Those layers were mapped in a projected spatial reference system (State Plane Coordinates), so we will create a GIS database for these

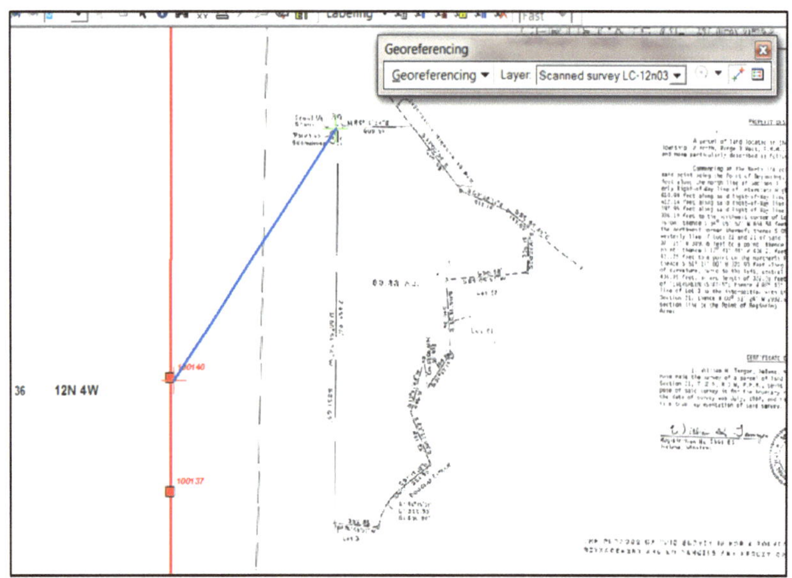

FIGURE 43 GEO-REFERENCING A SCANNED IMAGE

surveys using that same spatial reference. The GIS database that we create will have two attributes – one for the survey file number and one attribute for the scanned image file name.

Once the reference data are loaded into a map and we load our survey dataset we can then load scanned images of each survey, geo-reference each scanned image, then digitize the survey perimeter as a GIS feature, which we then attribute with the survey file number and the scanned image file name.

When we first load each scanned image, we might not see initially see it because the image does not have any geo-reference information associated with it. We therefore must geo-reference the survey scanned image. ArcMap provides tools to do this. We will geo-register the image by indicating where certain points on the image should be located in the map. For example, the survey may show a section corner that we may already have a point for in ArcMap in our PLSS point layer. We use the geo-registration tools to tell the software that this particular point on the survey image belongs at that particular point in our GIS map. We do this for enough points on the image to translate, scale, and rotate the image into the correct geographic location in ArcMap. Once that is done then we can begin digitizing the survey boundary.

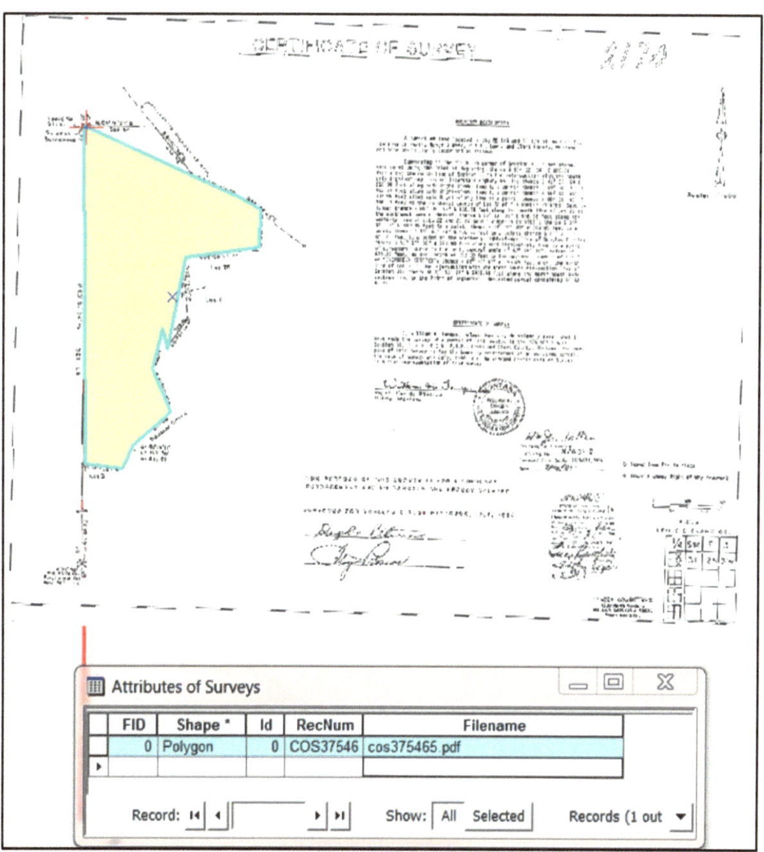

FIGURE 44 GIS SURVEY FEATURE DIGITIZED FROM SCANNED IMAGE

ArcMap has some geo-registration tools built in. The figure 44 shows the geo-referencing tool bar loaded in ArcMap. One of those tools allows us to fit the scanned file to fit the display (remember this is just a raster image). This works very quickly and helps to get image visible on the screen so that we can pick corners of the survey image to link to actual points on the reference layers. Once the raster image is visible on the screen, we can now link points on the image of the survey, to the PLSS points or parcel corners or whatever else that they represent. As each of these registration points is linked (image point to map feature) the raster image registration is performed on the fly. Bad points can be eliminated to achieve the best possible fit. However in this case, we are only using the scanned image to make an approximate perimeter for the limits of the survey, so the image registration need not be exact. It is just a visual tool for creating an index, so we do not need a very precise fit. Part of the problem that we have to deal with is that the reference GIS layers may have differing spatial accuracies, which may be less than our survey. In cases where we want to use the surveys to improve the spatial accuracy of the existing GIS layers, we would use more accurate geo-referencing and data conversion methods than this.

Once the scanned image is geo-registered closely enough for our purposes, we can begin editing the Survey_Index layer, to create a polygon feature for this survey. In ArcMap we start editing the survey layer and create a new polygon to represent the survey perimeter, by snapping the vertices (angle points) of the survey polygon to the angle points on the existing parcel boundary layer. When we have completed creating that polygon, we add the attribute values to the attribute table and save the changes.

This creates a polygon feature that represents the approximate extents of that survey. By having this as a polygon vector GIS layer, we will be able to perform spatial queries and analysis on that layer. Also, because we have added the file name of the scanned image of the source document as an attribute to the polygon, we can use that as a hyperlink. Hyperlink means that when we click on that survey polygon, the scanned image of the original document can be displayed on the screen. That way the user can read the image of the survey itself in order to find any other information that it may contain. This particular survey has information about some of the corners, senior lots, parcel reconfiguration and other interesting information.

As an example of the usefulness of a GIS layer, we'll perform a spatial query to get a list of owners of the properties that are within 100 feet of this survey. The figure 45 shows the spatial query window in ArcMap. Here are looking for the parcels that are within 100 feet of the selected survey. Those parcels are highlighted in yellow.

If these parcels have attribute information such as property owner name, mailing address and parcel size, we could use ArcMap's report engine to generate a contact list of the affected parcels. A single list could be generated, but the same information could be used to create a mailing list which could be fed into a form letter.

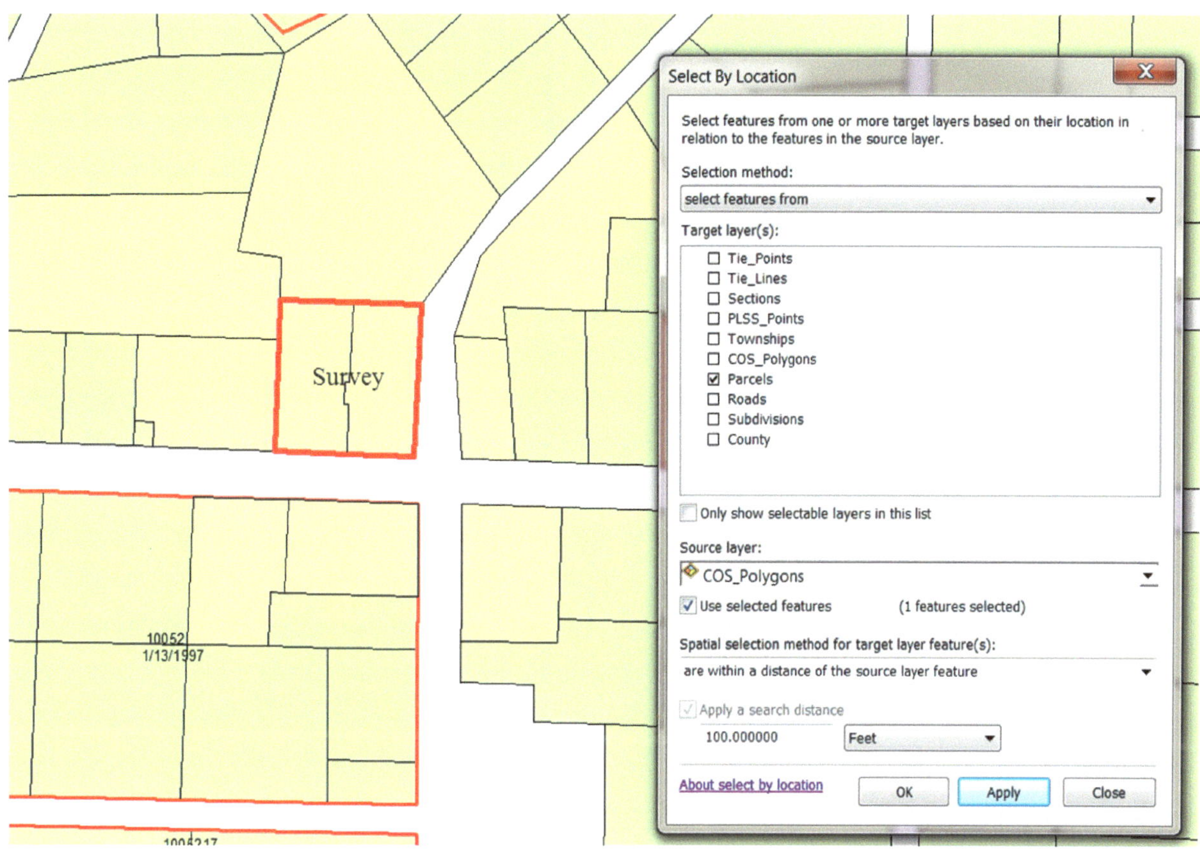

FIGURE 45 PARCELS WITHIN 100 FEET OF SURVEY (RED OUTLINE)

CHAPTER 18 CREATE A GIS DATABASE OF CORNER RECORDS

Some jurisdictions require that surveyors file a record of corner rehabilitation (or preservation) whenever a property corner is changed. These records are typically filed at county courthouse, or municipal office or county surveyor's office on paper forms, although some counties scan the forms into a PDF or graphic image. Searching the public records can be challenging for a number of reasons. The first issue is that for most jurisdictions, the surveyor must physically go to the public office to search for and make copies of the records. In small jurisdictions the public offices may open only part-time which limits access. Some counties are very geographically large and the courthouse or county surveyor's office may be many miles from the job site. Additionally some public offices put all the records into a cardboard box, some file the records in books sorted by location (township and range), some store the records chronologically by when the record was submitted, and some offices put such a low priority on the corner records that they may not know where to find them.

A solution for the problem of providing access to public records is to computerize those records and put them online. With today's technology, this is relatively simple. The most logical route is to use GIS in order to provide an additional means to search for records - a means based on location. Location based searches can be performed with web mapping tools and can also be made available using a smart-phone type of application which can tell a surveyor where the nearest corners are based on the surveyor's present location, and can provide

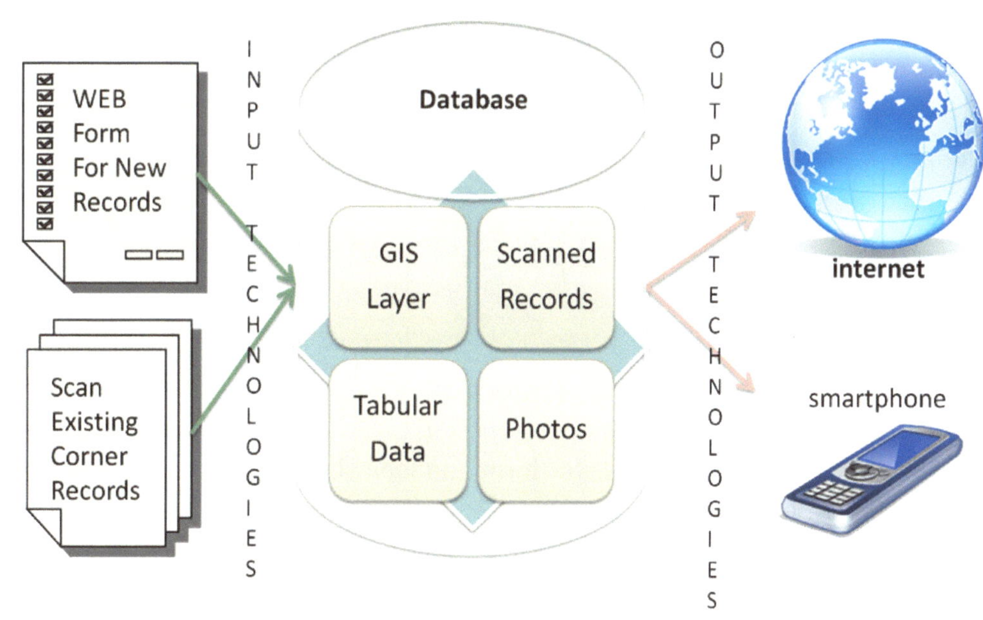

FIGURE 46 CORNER RECORDS SERVED VIA THE INTERNET

a link to the corner record document(s) any time of day, anywhere there is cell coverage. The technology solutions are feasible and developed relatively quickly. The main obstacles to achieving this are the cost and effort required to convert the paper documents to digital form and creating a searchable database for the scanned images, and the political resistance to change. Ideally, a statewide solution is the most logical in order to make the document form and the database consistent across a state. A statewide database also simplifies access for everyone. A statewide database ensures uniformity of corner records and the database, and makes it simple to develop an online form for submittal; an online submittal form saves the surveyor time, ensures uniformity of content and instantly makes the record public to everyone. Figure 46 shows a conceptual model of how a corner records database can be served via the internet.

An example of a statewide web application that provides a means to take in information from surveyors and store it a database that is served online is the Idaho-Montana Control Point Database application (http://gisservice.mt.gov/MCPDviewer). The application (figure 47) connects to each state's surveyor registration list to control who may create a Multi-state Control Point Database (MCPD) submittal account. The MCPD has an online control point submittal form that enforces content completeness and continuity. A database administrator, who is a licensed surveyor, then reviews the submittals for form and logical consistency. The administrator then publishes the data to the MCPD viewer. Using the MCPD viewer anyone can search for control points based on querying the database (e.g. by city, or by township, or by who submitted the points, etc.) or by querying the map by graphically selecting points or performing a search buffer.

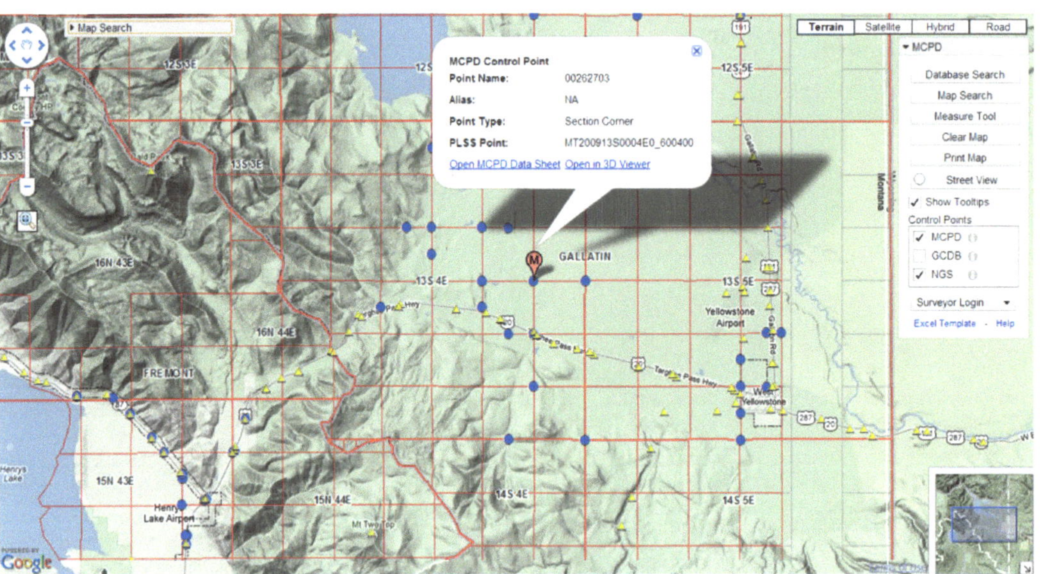

FIGURE 47 ONLINE CONTROL POINT DATABASE

The MCPD methods for collecting information from surveyors and publishing surveyor information can easily be applied to other types of surveying related information, such as corner records. There a couple of ways to create the GIS features that represent corner points. One method is to create a unique GIS point layer where each point represents a corner record and each point will contain associated attributes and a link to a scanned image. This method is the most universally applicable method and is the method to use when no other related GIS points exist.

OVERVIEW OF STEPS

1. Scan the documents into digital form - PDF is best because a single PDF file can contain multiple pages
2. Create a GIS point layer with at least one attribute for the scanned document file name.
3. Alternatively, one may add additional field to facilitate database queries (examples include township, range, section, date, surveyor name, corner name.

All existing corner records that are not already in digital form must be scanned and the digital files must be uniquely named. One file naming strategy is to name each scanned image file using the fully unique ID. Since there could be multiple records per corner, you should use a suffix of some kind to differentiate subsequent records. Adobe PDF (portable document format) format allows multiple pages per document thus organizing multi-page records, the image quality can be high, and the document can be zoomable. The Adobe Reader also gives the user a lot of control over how to print a document.

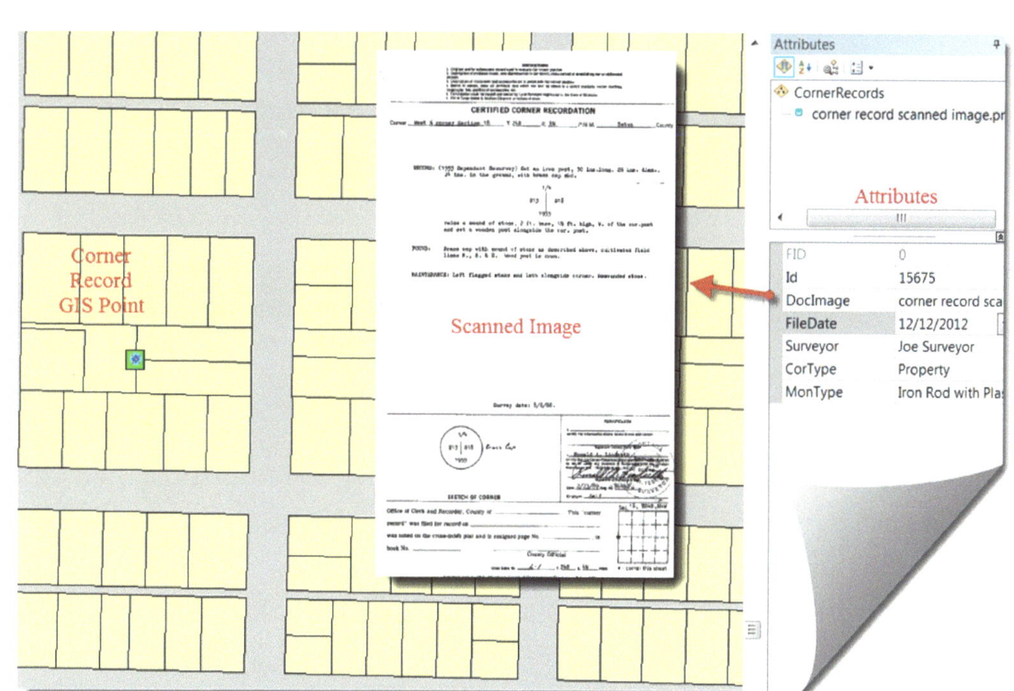

FIGURE 48 CREATING A CORNER RECORD GIS POINT WITH ATTRIBUTES

To create a GIS point for each corner record requires editing in GIS software - one can simple create a point for a corner record by clicking on the correct location on a map as shown in figure 48. Ideally, this should be done by snapping new points to any existing geometries of reference layers (such as parcel corners, or survey grid locations, etc.). This method is labor intensive - someone has to perform the GIS editing and ensure that the points are correctly placed on the map.

After each point is created the attribute values must be entered for each point. This should include the unique ID for the point as well as the file name of the scanned image of the corner record. The file name attribute will be used later to create a hyperlink to the corner record document. Note that every time that a new corner record is created someone must create the GIS point and add the attribute. This effort may be centralized at a responsible agency. Alternatively, the surveyor who creates the record could do this task automatically when he or she creates a new corner record by using a distributed web-editing environment.

WHEN A RELATED DATA SET ALREADY EXISTS:

When a GIS dataset of related point data already exists, it may be possible to leverage those existing points to represent the corner data. This can be done by joining or relating the corner point database to existing GIS dataset. For example in the western United States the Public Land Survey System has a point data called the Geographic Coordinate Database (GCDB). The GCDB contains GIS points for each PLSS corner. Because many property corners in the western US are on the GCDB, it is possible to use the GCDB geography to create a GIS layer of corner records. This only works when the corner records are on the same points as the existing (in this case GCDB) points. A database join takes advantage of preexisting geometry for the graphic and spatial location, and attaches additional tabular to certain point records based on a unique value in a tabular field.

Overview of steps

1. Scan the corner record documents into digital form as described above.
2. Create a simple database that has at least one field for the scanned document file name and another field for the GCDB point ID.
3. Join the corner records table to the GCDB point layer using the GCDB point ID as the join field.

The database table must contain a field for the fully unique GCDB ID and the values entered in that field for each corner record, must match the corresponding values in the GCDB point dataset exactly so that the software can perform the table join correctly. In order to recall the scanned corner record document, the database table should also contain a field for the scanned document image file name which can be used as a hyperlink to recall the scanned document file. Depending on how the system is setup it may also be necessary to include the file folder name as well. In addition to the two required fields, the database may also contain other fields as appropriate, such as recordation date, surveyor's name, etc. which may be help when performing database queries.

After building the database, each corner record document must be added to the database. In order to create a GIS layer the corner records database must be joined using a *table-join* to the GCDB point layer using the GCDB ID as the join field. Figure 49 shows an example table join in ArcMap using the GCDB ID field (labeled here as *Corner Point Identifier*) then retaining only those records that find a match. The result, as shown in Figure 49, is a geographical index of PLSS points (red squares) that have corner records (green square).

Creating a GIS layer for corner records simplifies database searches and spatial searches, and makes it possible to create maps that show which corners have corner records.

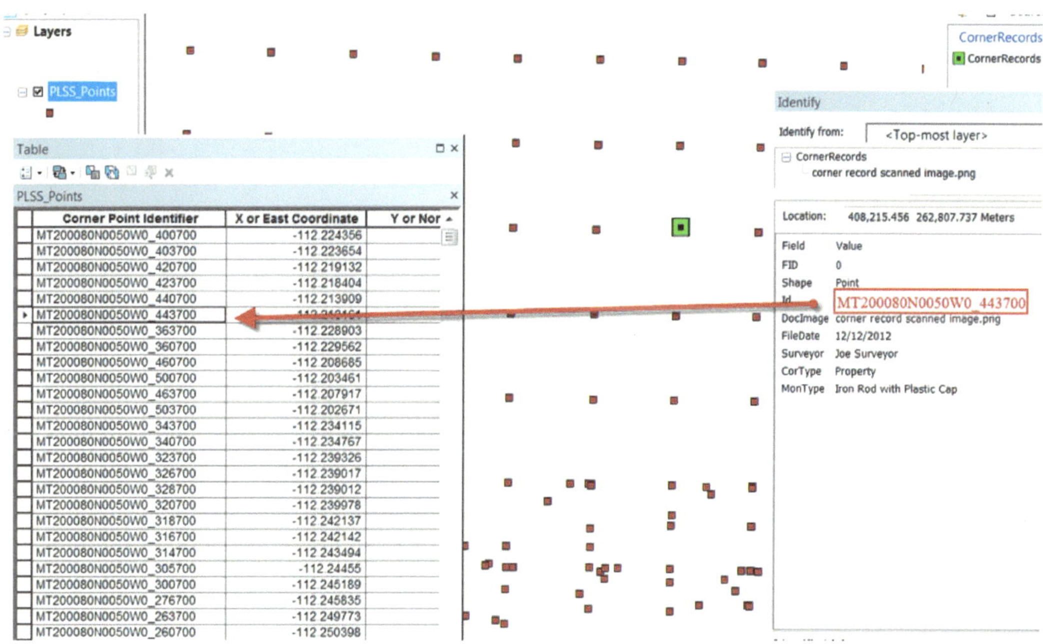

FIGURE 49 JOINING TABULAR DATA TO AN EXISTING GIS DATASET TO CREATE A GIS LAYER

CHAPTER 19 CREATE A GIS DATABASE FOR RIGHTS OF WAY & EASEMENTS

Rights of way and easements locations for roadways and utilities such as water lines, gas pipelines, electrical transmission lines and others can be mapped in GIS to be used as a planning tool, for right-of-way negotiations, understanding impacts, to aid in survey layout, to perform holdings inventories, and a many types of anlaysis.

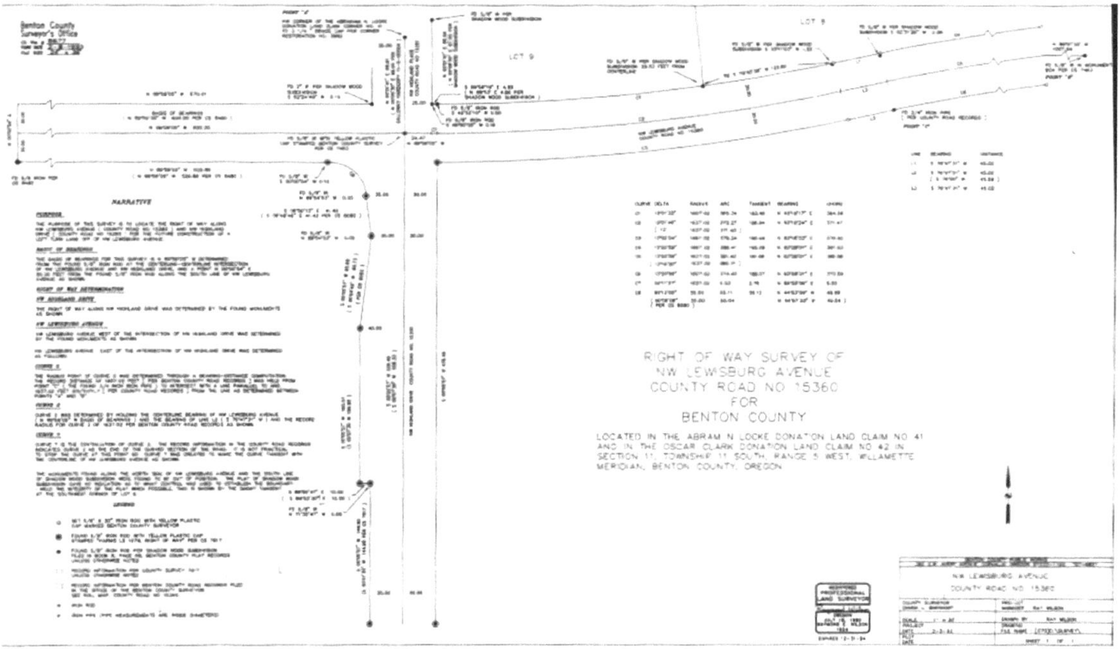

FIGURE 50 ROAD RIGHT OF WAY SCANNED DOCUMENT

Depending on one's objectives, schedule, and budget, one could use one of three methods to create a right-of-way GIS layer. The three methods discussed here are based on the GIS geometry used to represent the location of the right-of-way project. The method you use will be driven by how useful the product must be for you, how much time you have to create the product, and the cost to do the mapping. The choices of GIS geometries are Point, Line, and Polygon - any of which you may use to create a GIS index for the right-of-way. The chart below characterizes the time to develop the layer, the resulting spatial accuracy in terms of the geographical representation of the location and extents of the right-of-way, and the types of spatial analysis one could perform with the different types of geometry.

In all cases, any hard copy documents can be scanned and the scanned image (figure 50) can be hyperlinked to the GIS geometry. Additionally the scanned image might also be geo-referenced in GIS to display the scanned image in geographic space (figure 51). Geo-referencing the scanned documents will also aid in performing the GIS mapping – particularly for line and polygon geometries.

GEOMETRY	TIME TO DEVELOP	SPATIAL ACCURACY	CAN LINK TO SCANNED DOCUMENT	SPATIAL ANALYSIS
POINT	QUICKEST	LEAST	YES	COUNT, APPROXIMATE NEAR, APPROXIMATE DISTANCE
LINE	MODERATE	MODERATELY	YES	COUNT, NEAR, DISTANCE, INTERSECT
POLYGON	SLOWEST	MOST	YES	COUNT, NEAR, DISTANCE, INTERSECT, CONTAINS

If one's purpose is to create a quick graphical index that shows an approximate location of the right-of-way, then a point geometry will suffice. A point geometry can be quickly inventoried to show the total number of rights of way and their spatial distribution.

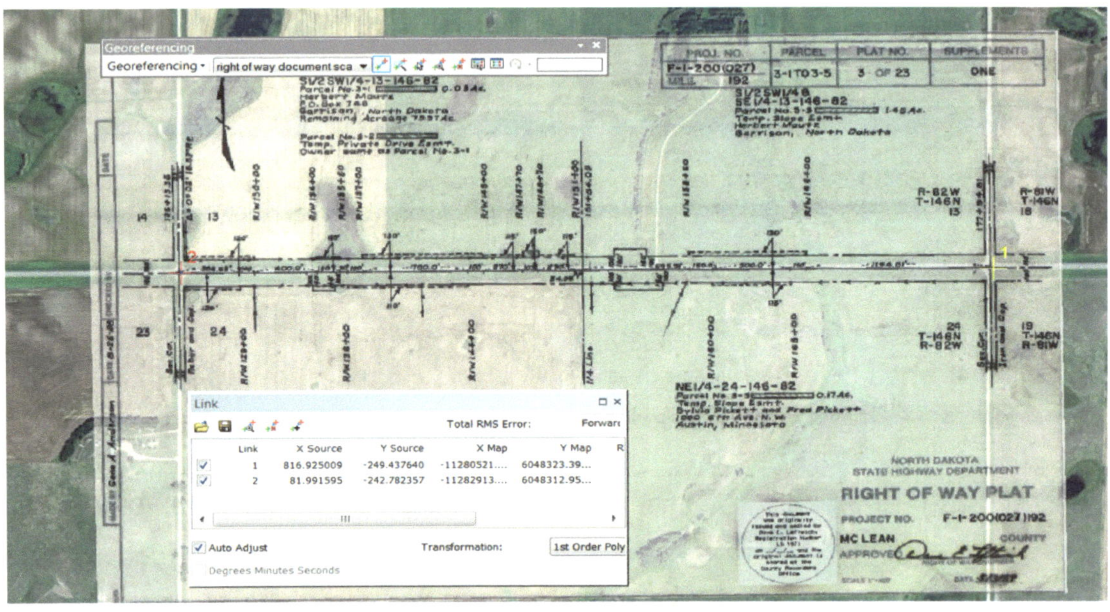

FIGURE 51 GEO-REFERENCING A SCANNED RIGHT OF WAY DOCUMENT

If one needs to visually represent the approximate linear extent of the right-of-way, then a line geometry would serve this purpose since a point geometry cannot. A line geometry can indicate the approximate location and linear extent and can support additional types of analysis such as determining what it cross, what is along it, what is near it - throughout its length.

To truly understand the footprint of the right-of-way, then a polygon geometry is necessary. A polygon geometry can represent the actual width and length and its variation along its entire length. In addition to the analysis that one can perform using the line geometry, the polygon geometry can be used to analyze impacts of the right-of-way footprint – such as determining the actual area (acres or hectares) of the right-of-way.

Descriptions follow below of each of these three ways to graphically represent right-of-way in GIS.

POINT

The simplest and quickest way to map a right-of-way project is to create a GIS point for the location of the project (figure 52), then associate the right-of-way project documents to the point. The associated documents could be scanned images of plans, or CAD drawings, PDFs of conveyance documents, etc. The project documents may be stored independently of the right-of-way project GIS data and a table or database would store the linkages between the two sets of data. This can be done as attributes in the GIS dataset that contain the file name to the documents, or a separate table that relates the records in the GIS file to the record or records in the documents database.

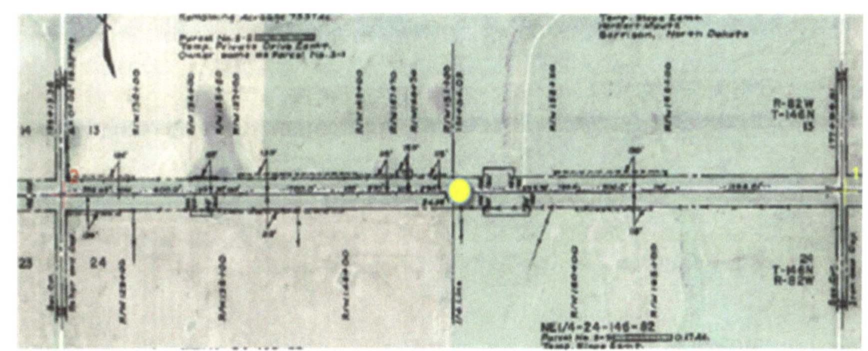

FIGURE 52 REPRESENTING RIGHT OF WAY LOCATION WITH A POINT

The advantage to creating a GIS point is that they can be created very quickly and easily, and thus inexpensively. The points may represent the middle of the project, or the starting point or end point of the project. The disadvantage to using point geometry for a GIS index of right-of-way is that a point will not represent the full extent, length, width and geometry of the right-of-way. For display purposes, a point will not show what the right-of-way looks likes. Additionally the ability to perform spatial analysis is greatly diminished. For example: determining whether a certain condition (such as wildlife habitat) falls within the right-of-way cannot be done reliably with a point. Although, it is possible to perform buffer analysis using a point, such as whether or not a condition is within a certain distance of a project, the buffer is created around the point and not the full extents of the right-of-way.

Usually, spatial analysis of a point which represents something that may have a large and complex shape can only a gross estimate, at best.

LINE

Right-of-way projects can be mapped using a line (figure 53) in much the same way as a point can. The line however, can represent the location *and* the linear extent. The line representations typically map the centerline geometry of a right-of-way project. Just as with a point, the line graphic may also have attributes which can be linked to source documents and other tabular information.

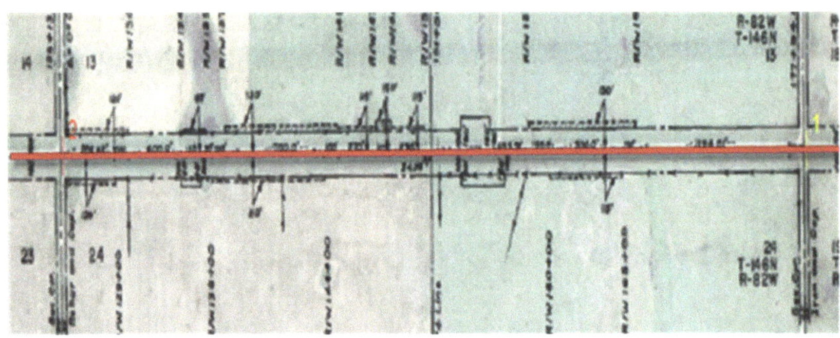

FIGURE 53 REPRESENTING RIGHT OF WAY LOCATION WITH A LINE

Using a GIS line to create a right-of-way index has advantages over using a point in that the full linear extent and possibly the linear shape of the right-of-way can be drawn for display, and to allow improved spatial analysis. With a line, spatial analysis such as *within distance of* is improved by the ability to search from either end and anywhere along the line of the project. With the line geometry, it is easy to generate a buffer of the full width of the right-of-way. Also, with a line to represent the right-of-way, it is possible to analyze whether something crosses the right-of-way or whether the right-of-way crosses something.

The disadvantage to mapping lines as opposed to points is that creating the line takes more time. The spatial accuracy of the line will vary with the method used to draw the line. If the line mapped is generated by sketching it will not be as accurate as when it is drawn using the coordinate geometry (bearings and distances of the centerline for example). However, if the geometry already exists in a digital form such as CAD, then creating the line can be as straight forward as converting CAD geometry to GIS features.

POLYGON

For greatest accuracy which will enable the most utility, a GIS index of right-of-way should be mapped as polygons (figure 54) that represent the entire geometry. Polygons can also be created by sketching, tracing the scanned image, or by COGOing in the right-of-way bearings and distances. A polygon can best represent complex geometry, extents, changing width, and location of the right way. Polygons will provide the best visual representation, and affords the most accurate spatial analysis.

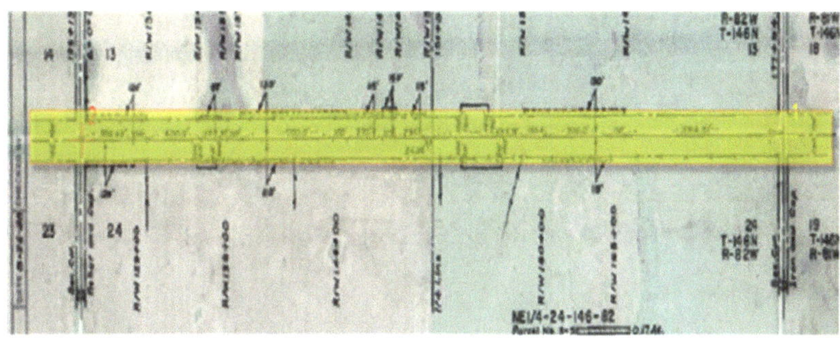

FIGURE 54 REPRESENTING RIGHT OF WAY LOCATION WITH A POLYGON

The disadvantage to creating the polygons is the time that it takes to digitize in the complex geometry. However, if the geometry is already represented in digital form such as a CAD drawing, then the entire process can happen far more quickly and with less likelihood of mistakes in data entry.

Complex geometry can be generalized to a simple rectangle to show the length and approximate with of the project as shown below in the Minnesota Department of Transportation right-of-way index map (figure 55).

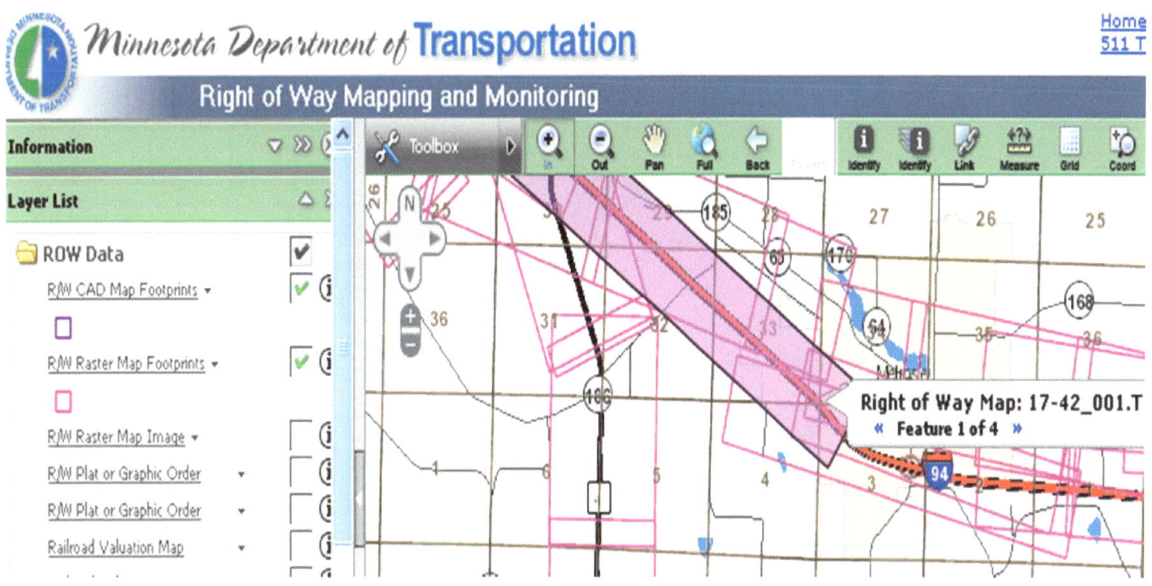

FIGURE 55 MINNESOTA DEPARTMENT OF TRANSPORTATION ONLINE RIGHT OF WAY ACCESS

When mapping rights of way and easements, one may chose the method that best suits the objectives for the intended use. However, all three methods (geometric representations) can be used in a phased approach – starting with the simplest geometry to generate a quick index, then later creating the more complex geometries as time and budgets allow. Additionally, when two or three of the geometries are available then they can used for *scale dependent* mapping applications such as web maps. The different geometries can have scale dependencies so that, for example at small scales the points appear, at large scales the lines appear, then at very large scale the polygons appear.

CHAPTER 20 GOOGLE EARTH AND SURVEYING

Nothing has so radically transformed GIS as the 2005 release of Google Earth ™. Google Earth combines a very simple yet powerful interface with stunning graphics and a planet full of data to which anyone may add their own information and share that information with the rest of the world in just a few short clicks. Because of its simplicity and power, Google Earth has become widely adopted in a relatively short time. It is simple to use, fun, and stable.

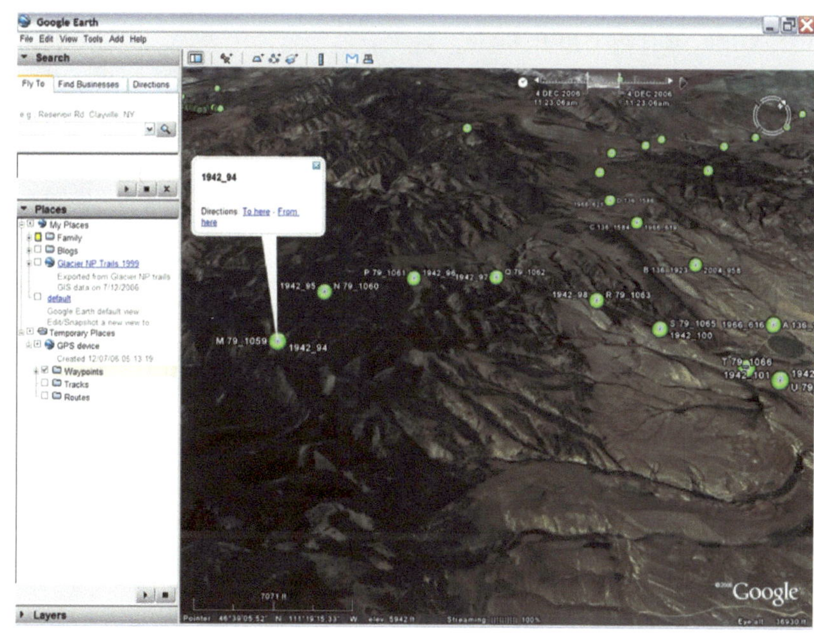

FIGURE 56 NGS POINTS IN GOOGLE EARTH

Google Earth ™ has revolutionized GIS by providing a free and simple to use GIS application that anyone can use. Requiring only a good internet connection and a 3D capable graphics card, Google Earth provides global aerial imagery at varying scales. Users may download the Google Earth application, which is free for personal use, $400 for professional or business use. Google Earth works by connecting to the internet to stream data to the users. The user has a few simple tools to control navigation such as pan, zoom, and tilt (for 3D perspective). Other features allow the user to enter an address in a variety of formats (street address, or coordinates) or name of a place and then zoom to that location.

Additionally users can add their own geo-referenced data to display in Google Earth. The required format is Keyhole Markup Language (KML), or GPX (GPS exchange). KML is a simple text formatted syntax that provides basic information regarding location, data type (points, lines, and polygons), symbology, and attributes. Many free utilities will convert data from one type or another into a KML format for use in Google Earth. Surveyors may be interested in converting data from AutoCAD, MicroStation, or ASCII formats into KML. Autodesk provides a free utility to convert AutoCAD data (from certain versions). Bentley Systems Inc. provides a utility for its SELECT subscribers. One need only search the internet for a utility to convert from your favorite format into KML or GPX format suitable for Google Earth.

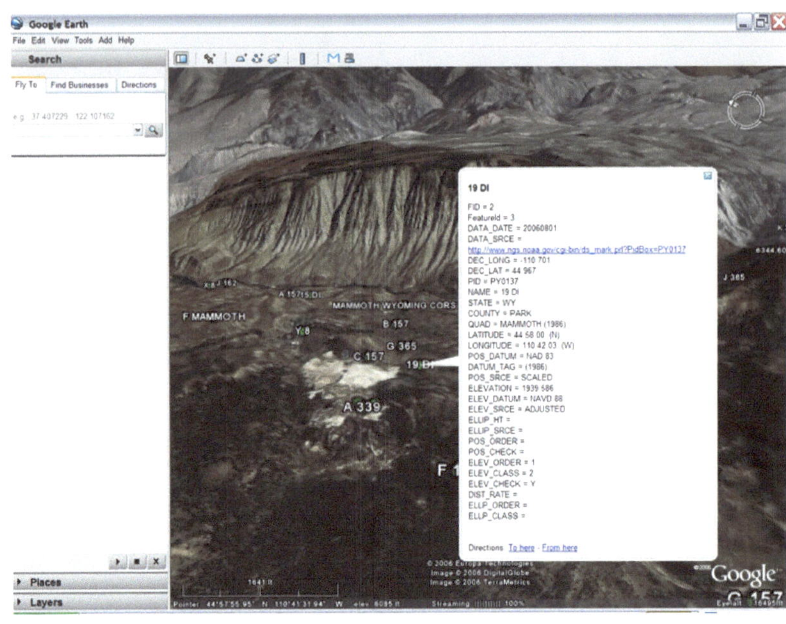

FIGURE 57 NGS POINTS KML FORMAT IN GOOGLE EARTH

GPX format is compatible with some GPS devices and uses XML schema for describing waypoints, tracks and routes. Google Earth will accept GPX data. For example, one may easily convert NGS datasheets (DAT) with very simple converter to GPX format. In this way the location of NGS points may be displayed in Google Earth (figures 56 and 57). Note that Google Earth provides Directions: To Here/ From Here option within the point feature balloon. This handy tool will provide driving directions on how to get to or from the point. GPX is a simple format for integrating GPS compatible data that will display in Google Earth. Additionally there are free utilities exist that will convert KML format data to GPX format, which one may then upload to a GPS device. With the data loaded into a GPS device one can then navigate to the point.

I opt for a more robust solution to representing NGS points in Google Earth. I find that integrating my own data with Google Earth is very simple to do, and is a powerful tool for providing context and location that are helpful for planning projects. In the following example, I extracted National Geodetic Survey (NGS) data for Yellowstone National Park, in order to see where geodetic control was available within the park. To do this, I downloaded the NGS points for an area in shapefile format, which I then loaded into ArcMap which has a utility to export to KML format to show in Google

Earth. Once I had created the KML file, I merely drag and drop it into Google Earth. Google Earth then automatically zooms to the spatial extents of the points that I loaded.

Within Google Earth, I can click on a point to list the attributes of the point. If I had included the DATA_SRCE attribute, then I can also click on that attribute which is a hyperlink to the NGS datasheet. That sends a request to the NGS server to pull up the data sheet for that particular point. (Figure 57). This is a very nice feature because the KML files are typically compact and small enough to email, yet when I want the full datasheet, I can readily get it, and more importantly, when I click on the hyperlink I get the latest datasheet. With the NGS points in Google Earth, I can see where the points are located, how to access them, whether or not they are likely to be open to the sky or traffic issues etc. In addition, I can zoom around the points and look at them in perspective view or see from them from any angle. Although this is more complicated path to getting NGS points into Google Earth, I find that the end result provides me with a more useful solution. The ability to hyperlink to the latest and most up to date datasheet is important for obtaining the more recent information on recovery, coordinate adjustments, and any other information that may change.

In addition to points, Google Earth will also take lines and polygon overlays. Polygons are useful for features that cover an area, and the polygon itself can by symbolized with a solid color, or a transparency. One may, for example convert zoning to a polygon KML feature then overlaid in Google Earth (figure 58 City of Portland, Oregon USA). An agency or company could communicate project information such as the location of sewer and water lines for a development project using Google Earth.

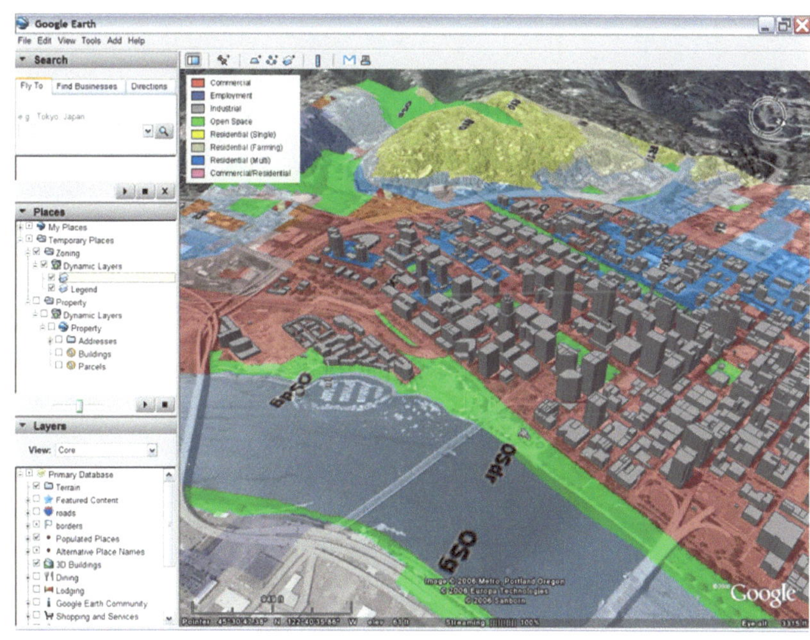

FIGURE 58 PORTLAND OREGON ZONING OVERLAY IN GOOGLE EARTH

Because KML are usually pretty small they can easily be emailed, or they can be shared by posting them on a website for others to access. This means that sharing your GIS data with a client or colleague is very easy to do. Google Earth is a rich and powerful tool that makes it easy to use GIS and to share GIS information with others. Because of this, Google Earth is putting GIS in everyone's hands.

CHAPTER 21 CREATE A GIS MAP OF YOUR SURVEY CONTROL WITH GOOGLE FUSION TABLES

Here we discuss a simple and quick method for creating a GIS map of survey control points, using Google's fusion tables mapping application. You must have a spreadsheet or text file (as comma separated, or other text-delimited format) which includes a field or field of geographic coordinates for your points. There is a file size of 100 MB.

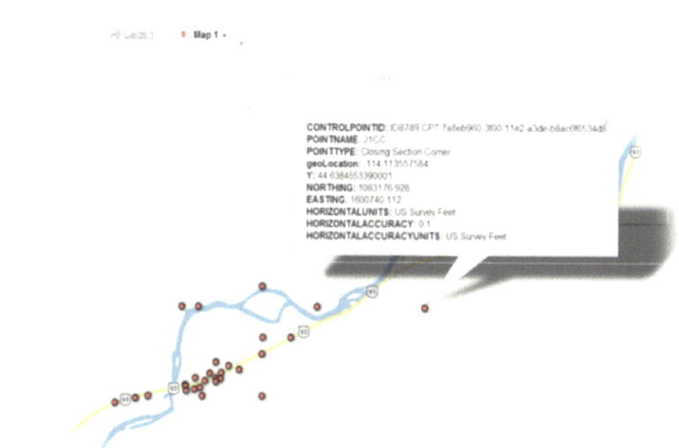

We will start with an Excel spreadsheet file of control points (perhaps exported from a surveying software program). The control point spreadsheet file contains information about the control points, and includes one or two fields that have the coordinate information for the location of the control point. We will load the spreadsheet file into Google's Fusion Tables, and then identify the field or fields that contain the coordinate information. Once that is done we can very quickly make a map of the location of each of the control points. This can all be done in less than 20 minutes (if one already has a Google account).

Google's Fusion Tables are available for free, to anyone who has a Google account. You can create a table and load your existing data (in spreadsheet, text delimited, or Keyhole Markup Language [KML] format). Once you create a table you add more data to it, edit existing records, filter the data, and map it. Your Fusion Table resides in the Google Cloud, and you may keep it private, share with selected individuals, or share it with the world.

As part of the process for creating a Fusion Table in Google, we import our spreadsheet and identify which row in the spreadsheet contains the headers as shown in figure 59. Note that the spreadsheet must contain a header row in order to name each field (also called column) of data. The data from the spreadsheet are now imported to a Fusion Table – shown in figure 60.

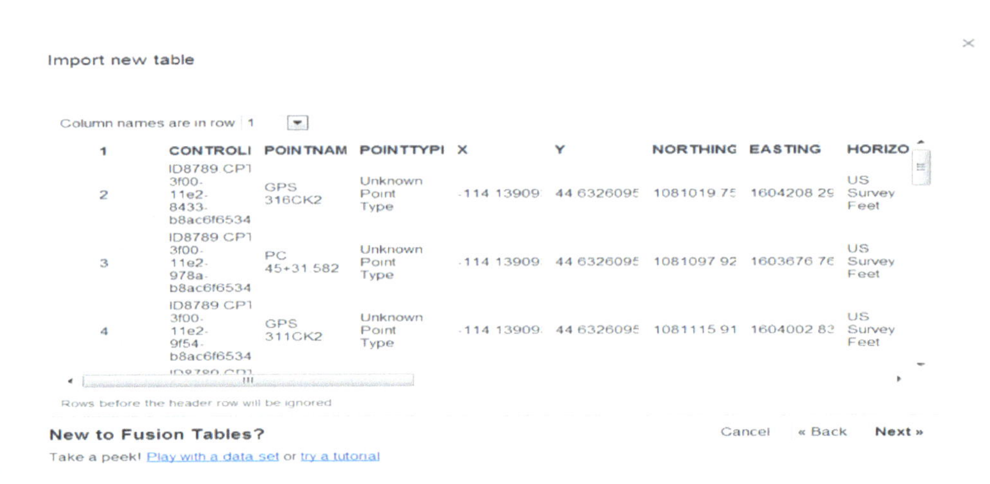

FIGURE 59 CONTROL POINT FILE TO UPLOAD TO A FUSION TABLE

Note the various tabs available on the Fusion Table (figure 60). These tabs provide various functions such as how the data are viewed. For example, shown here the data view is a list of rows (each record is a row). Also available are Tools and File operations and some editing options.

One can click on the column headers to sort the records in ascending or descending order for that column.

The data map be filtered or summarized for different perspectives on the data. Numeric fields may be charted to show a graphical perspective on the data.

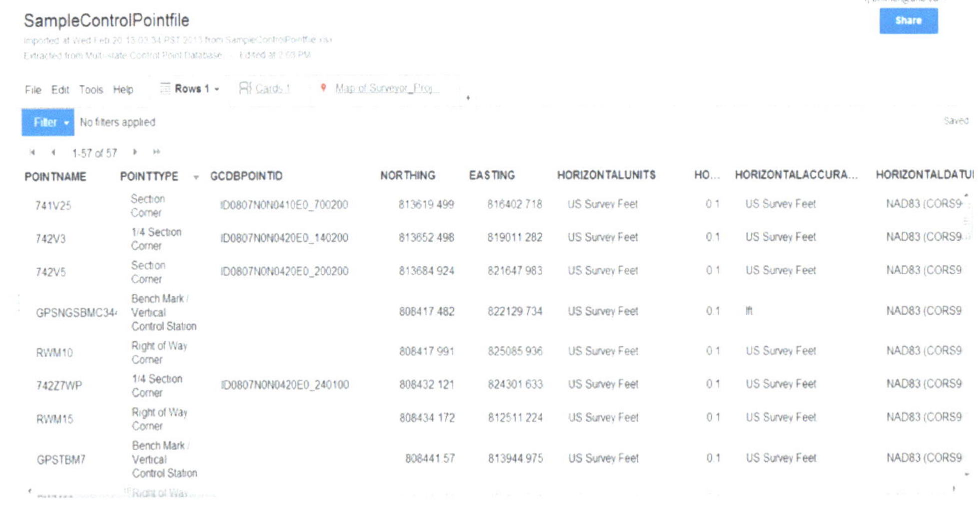

FIGURE 60 CONTROL POINT FILE DATA AS A FUSION TABLE

Figure 61 show a summary of these data for the MonumentType field. This operation summarized the data by each of the monument types that it found in this database.

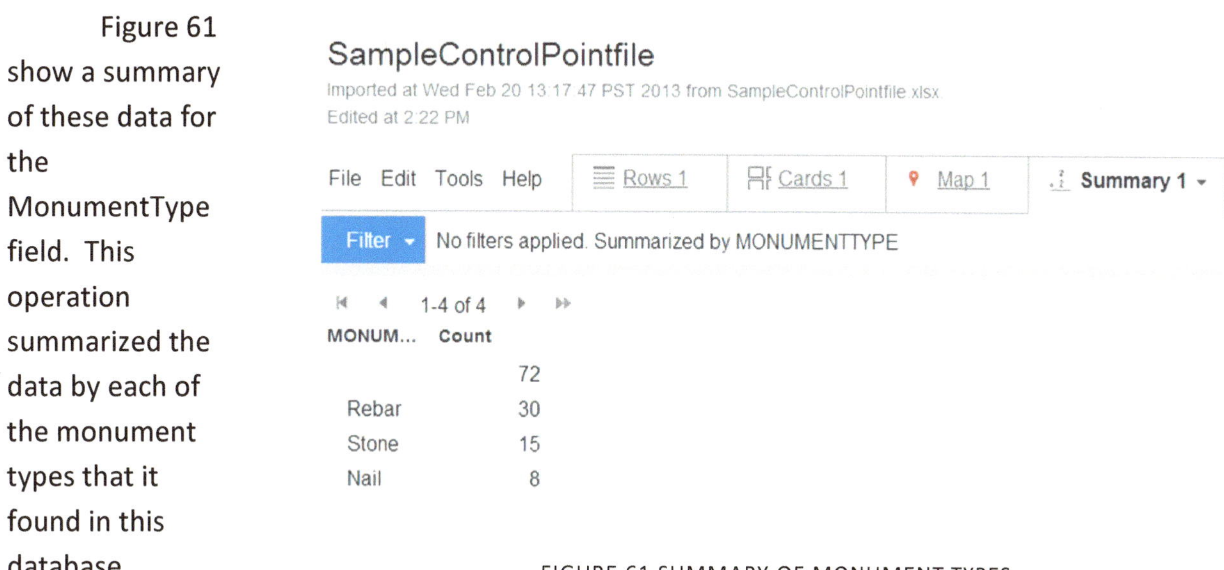

FIGURE 61 SUMMARY OF MONUMENT TYPES

Once the spreadsheet's data are loaded into the Fusion Table, we must identify which fields contain the location (coordinates) information. In this data set, the coordinates we will use to for mapping purposes are the X and Y fields which contain geographic coordinates for eastings and northings in decimal degrees format.

Once the coordinate data have been identified, simply click on the Map tab to create a Google Map of your control points like the one shown in figure 62. You can zoom in and out, pan around and change the background maps.

FIGURE 62 MAP OF CONTROL POINTS

You can share your map and/or your data in any number of ways – email, social media, provide access to your map for co-workers or clients, or the world.

CHAPTER 22 DESIGNING A MAPPING CONTROL NETWORK FOR GIS

Here we discuss a way to design mapping control for GIS by using a case study for a local government in a state in the western US. This case study – Lewis & Clark County, Montana (figure 63) began its GIS development in the late 1990's during the same time that the state of Montana embarked on a state-wide parcel mapping project. Lewis & Clark County cooperated with the state on the parcel mapping for Lewis & Clark County so that the county's parcel mapping and the state parcel mapping were one. Prior to the parcel mapping effort, Lewis & Clark County created a mapping control network to support the parcel mapping and other GIS mapping that the county planned to do. This case study discusses the planning and execution of the mapping control project.

FIGURE 63 CASE STUDY AREA LEWIS & CLARK COUNTY MONTANA USA

Lewis & Clark County is situated in central Montana on the eastern slopes of the Rocky Mountains and encompasses approximately 3,500 square miles (more than 906,000 hectares). The county stretches over 100 miles (161 km) from the Helena Valley in the south, to the Sun River in the north, and over 60 miles (97 km) from the Rocky Mountain divide in the west to the Big Belt Mountains to the east. It has a population of about 53,000 people with approximately half of that number in the state capital and county seat in the City of Helena at the south end of the county. The county has three federally designated wilderness areas, three state wildlife management areas, four national forests and many other federal and state lands. Of the 2.2 million acres (890,000 hectares) in Lewis & Clark County, parcel sizes range from large ranches of thousands of acres (hundreds of hectares) to small urban lots.

The vast distances, coupled with the concentration of population (and thus mappable human activity) in a very small portion at the southern end of the county, created some interesting mapping challenges for GIS. The Lewis & Clark County GIS Implementation Plan of 1998 identified more than 50 datasets to map for GIS, including parcels, aerial photography, roads, city boundaries, zoning, wildlife habitat, geology, building footprints, and others. Because of the large number of datasets and the variety of map scales needed, integrating them spatially required good survey control. The vast distances magnify discrepancies when maps are brought together from a variety of sources, or mapping is done from differing points of beginning, or the mapping is done at differing scales, or using differing basis of bearings or map projections, and differing levels of detail and spatial accuracy. To bring order to this array of map data, mapping control is essential. Map control is the framework upon which all the datasets must be registered for proper geographic placement, scale and alignment. Mapping control also provides checkpoints for validating coordinate values of map sets.

CONTROL PROJECT ORIGINS

Lewis & Clark County began developing a Geographic Information System approximately in 1998. A GIS needs assessment study performed for the Lewis and Clark County identified the importance of the parcel layer for the county as well as state agencies. Prior to the statewide mapping project, the county had begun converting some parcel maps into digital format using coordinate geometry (COGO) methods. The original parcel maps were hand drafted assessor maps on polyester film. For the initial county parcel mapping, the assessor maps were scanned and geo-registered to an existing 1:100,000 Public Lands Survey System (PLSS) GIS layer.

The state parcel mapping project, however, used the Geographic Coordinate Database (GCDB) created by the US Bureau of Land Management (BLM) for mapping control. The GCDB is an improved PLSS reference frame that provides superior control for parcel mapping and other GIS projects. However, the GCDB for Lewis & Clark County had not yet been built at the time the county and state started the parcel mapping project. The county recognized the need for a cohesive network of mapping control on the PLSS that would support the parcel mapping for the state and the county. The county also recognized the value that a control network would have for other agencies in the area, such as the BLM, US Forest Service, the Montana Department of Transportation, and private surveyors. Therefore, the county brought together the agencies that were interested in creating a control network, in order to discuss their requirements for such a project. Representatives of private and government agencies met to express their interests in control and to describe the nature of the control they needed.

The team of public and private sector surveyors and GIS professionals identified a variety of surveying and mapping control requirements to suit a variety of current and future projects. Each of these various needs had differing accuracy requirements that fell into three tiers of control.

TIER I CONTROL

Tier I control was high order control for any future surveying and mapping projects, such as control network extension and densification. The Tier I points established in the City of Helena was used to adjust the city's existing local geodetic network into the North American 1983/91 horizontal datum.

The city had records the survey ties to all its control points, so it was be able to use the high accuracy points to control the adjustment. This brought that existing control up to NAD 83/91. All Tier I control points were blue-booked and incorporated into the national framework of Community Base Network (CBN) maintained by the National Geodetic Survey.

TIER II CONTROL

Tier II control was the most prolific set of control, has an accuracy sufficient to meet the greatest variety of mapping needs, and is based on public lands survey system corners (PLSS – section and township corners). The Tier II control was the most significant control for the parcel mapping project because the Tier II coordinates derived for the PLSS were submitted to the Bureau of Land Management to control the GCDB, thereby significantly reducing the error estimates of the GCDB in the areas of control. More importantly for Lewis & Clark County was that the Montana Department of Administration used the improved GCDB to map the aliquot parts parcels (about 11% of the parcels, or 70% of the area) of Lewis & Clark County. The county also used the GCDB to control the other methods of parcel mapping within the county (on-screen digitizing and Coordinate Geometric methods). The Tier II control on the PLSS could also be used for photo control wherever those PLSS points are identifiable in a photo. Future photo projects would require pre-marking the township or section corners but wouldn't require any additional surveying in those areas.

TIER III CONTROL

Tier III control was project-oriented control developed to meet a limited specific need. This control was of lower accuracy (sub-meter typically) and obtained "on-the-fly". That is, a coordinate could even be developed for a non-monumented point such as a road centerline, or an unstable monument such as a lot corner in a subdivision.

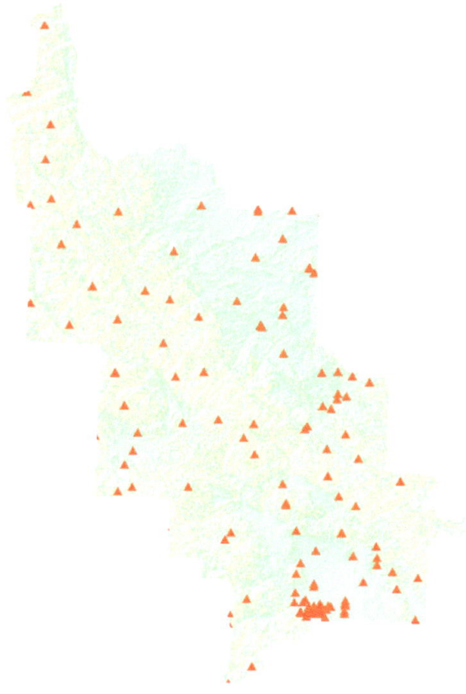

FIGURE 64 EXISTING HORIZONTAL CONTROL

THE SURVEY PLAN

Although there was some existing vertical and horizontal control throughout the county as shown in figure 64, it did not adequately meet the criteria or the design requirements partly because the existing control was located on mountain tops and thus not readily accessible. Additionally, because the control was located at arbitrary positions that were not related to GIS features they could not be used for GIS mapping. In fact, about one-third of the existing 1st order points were located in wilderness areas – far from any roads or things that needed to be mapped. Since the advent of Global Positioning (GNSS), survey control points no longer needed to be intervisible, so they don't need to be placed on mountain tops. Control points may now be placed nearly anywhere that the sky is visible. For the Lewis & Clark County project they looked to place control points on PLSS corners because the parcel mapping is based on the PLSS. They were also interested in locations that were easily accessible - such as near a road.

To meet the needs of the various uses for horizontal and vertical control, the survey project planning group agreed that additional control should be added. It also agreed that the control should be three tiers, based on the accuracy requirements. Tier I, high accuracy control, was needed to lock all subsequent control onto the national geodetic control framework (Federal Base Network, or FBN). Tier II, a centimeter accuracy level control on the PLSS, to provide a dispersed framework for parcel mapping and meet present and future control needs for the public and private sectors. Tier III, non-permanent "spot" control, may have any level of accuracy but probably decimeter level. Tier III control was project specific such as subdivision corners that help control a particular parcel map, and thus Tier III points were collected on an as-needed basis and not part of the survey control network.

CONTROL NETWORK IMPLEMENTATION PLAN, TIER I (HIGH ACCURACY)

To supplement the high precision controls of the National Geodetic Framework, including three existing B-order High Accuracy Reference Network (HARN) points, new high accuracy points were added for the county, and four existing medium accuracy points in the City of Helena control network were upgraded to A-order as shown in figure 65. The county and city points became part of the Community Base Network. These points were surveyed to an accuracy of one part in ten million (1:10,000,000) with two centimeter accuracy on the horizontal and vertical components. The county's intention was to distribute the new points out far enough so that the high order control completely surrounded the parts of the county mapping areas. That constrain all subsequent control within a known bounded perimeter, thereby constraining errors inside know values.

TIER I NETWORK GEOMETRY

Mapping control must be on the outside of the mapping project in order to contain all errors within the project area. Therefore the high accuracy control needed to be spread out far enough to completely contain the privately owned lands within the county. The control that existed prior to this project was sufficient to contain only the Helena Valley and none of the other populated areas. That meant more control had to be placed on the far east of the southern end of the county, the far northwest, and a point at the eastern edge at about the central part of the county (figure 65).

TIER I MONUMENTATION

Because these high accuracy points are important monuments, and because of the large amount of work and the high expense needed to generate their coordinate values, these points had to be made of durable material and located in a place free of local disturbance (natural or man-made).

TIER I ACCESSIBILITY

To ensure legal access to the points, and to minimize travel time to get to the point with equipment, these points were placed in public rights of way where a vehicle can be safely parked nearby. These points are all easily accessible by car or truck.

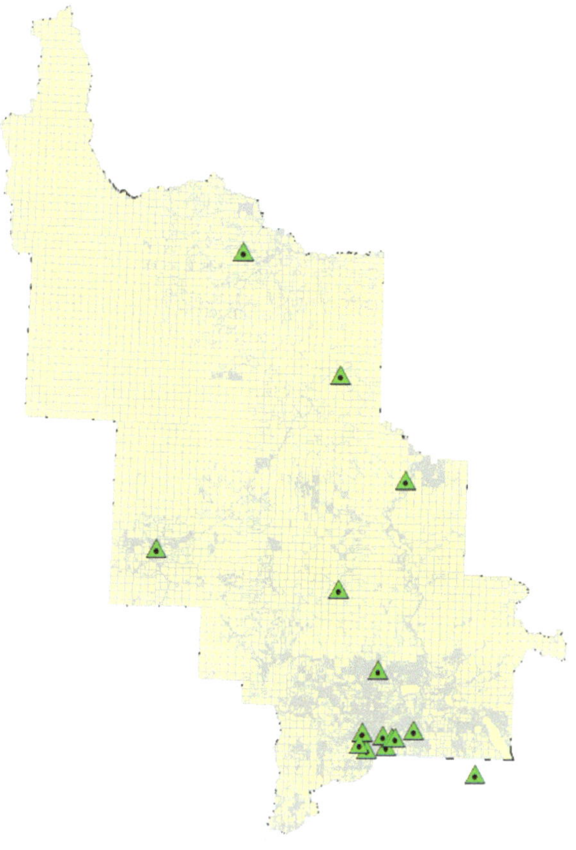

FIGURE 65 NEW HIGH ACCURACY CONTROL NETWORK (TRIANGLES) – GRAY LINES ARE PARCEL BOUNDARIES.

TIER II (CENTIMETER ACCURACY LEVEL CONTROL)

The second tier of control is placed on the PLSS. These control points don't require the level of accuracy of the high accuracy points. Here a few centimeters to a few decimeters will suffice for horizontal and vertical coordinates. These points will be tied into the high accuracy network and adjusted. This procedure will provide the consistency needed throughout the county. This means, for example, parcel mapping that starts in the southern part of the county and works north should match mapping started in the north and working south. There should be no breaks or discontinuities (equation stations) when map projects come together. The first phase of this project is to create the initial framework on the PLSS. Later Tier II control will be added to densify the control, primarily in the Helena Valley and other areas where needed.

TIER II NETWORK GEOMETRY

Since Tier II control for this project is on the PLSS, network geometry is dictated by the availability of stable monuments on township and section corners that are visible to the sky for GPS observations. The selection of which PLSS corners to use was based on the desire to control lines of the PLSS to contain the parcel mapping. The first choice for PLSS positions is township corners, but for a variety of reasons (mountainous terrain, agriculture, road building, etc.) not all the desired township corners exist. Nor are all the township corners useable for GPS observations. Thirty-eight "optimum" PLSS corners were selected, with the recognition that the first choice may not have a stable monument or be accessible or available for GPS observation. A private surveying firm would be contracted to research the 38 PLSS corners for stable monuments and to determine their suitability for GPS observations.

TIER II PERMANENCY

Due to time and cost considerations it was decided not to re-monument any PLSS corners for this project. Since the intention for the Tier II control was to develop horizontal coordinates on the PLSS for parcel mapping, the team decided to accept any durable PLSS monument, whether it was a brass cap on a three-inch pipe or an iron rod. Chisel marks in rock outcroppings were acceptable, but monuments that did not have a clearly defined point (such as a stone mound) were not.

TIER II ACCESSIBILITY

Every effort was made to select points that were not more than one-quarter mile from a public road and which were on public lands or rights-of-way. Accessibility is the important feature.

TIER II SPACING

Township corners were the objective of the first phase of Tier II control. Therefore, points were established about six miles apart. Since the focus, however, was to put Tier II control around the populated areas, not every township corner in the county was assigned a point.

PROJECT PRODUCTS

The high accuracy points will be blue-booked and become part of the CBN. Both the high accuracy points and PLSS coordinate values will be filed at the county Clerk and Recorder's office as a certificate of survey (Montana does not have a filing requirement for geodetic control). The control point information will also be entered into a control point database in the Lewis & Clark County GIS office. This will put the coordinate values into the public domain to be available for any mapping or surveying project.

The coordinate values on the PLSS will be submitted to the BLM for the GCDB that will then come back to the county and be converted to a GIS line coverage.

Tier I control were co-observed by the city and county as part of the 1999 NGS Re-observation Campaign. The data that the city and county collected were submitted to NGS and was processed along with other NGS data. Those high order control points are now part of the NGS control network. Thirty-eight section and township corners were coordinated using GPS observations as Tier II control. The resultant coordinates were submitted to the BLM where they were incorporated into the GCDB as control points. These control points are maintained by the county.

CHAPTER 23 USE GIS TO PLAN A SURVEY CONTROL NETWORK

GIS is a very useful planning tool where pertinent data are available. GIS provides analysis tools and visualization tools that help one to understand the spatial relationships between existing survey control, the areas where mapping will be done, and the physical, cultural, and built environments. With GIS we can analyze these spatial relationships and answer questions we may have such as where are the holes in our network, where do we have legal and physical access to property and a view to the sky, etc.

Here we consider in-filling an existing control network of Continuously Operating Reference System (CORS) stations in the state of Montana. Montana is a geographically large state with a small and dispersed population. There is an existing CORS network; however the existing network has many gaps in it. We will use GIS to plan where to place new stations. To help plan the placement we will consider the following criteria:

1. Existing control location and type: CORS stations
2. Existing control spacing : minimum of 70 KM spacing
3. Terrain – slope analysis: on flat terrain (< 5% slope)
4. Legal considerations - right to access: must be on public lands
5. Prioritize by need – Population: more populated areas get higher priority; areas with no people get lowest priority.
6. Located within 5 km of a town.

GIS analysis can answer each of these questions individually and the answers can all be brought together to provide a map of optimum locations for placing new stations. We wish to rank the areas where there is no coverage, based on meeting the above criteria, then identify the highest ranked town in each zone.

23.1 EXISTING CORS STATIONS & SPACING

We begin by making a map of Montana showing existing CORS stations downloaded from the US National Geodetic Survey website. There exist a few dozen stations in and surrounding Montana.

Based on NGS guidelines the desired minimum spacing between stations is 70 km. We analyze the spacing of the current locations by creating a 70 km buffer around each station point. This is a GIS analysis operation, which we perform in a few minutes. Here we dissolve the areas of overlap of adjacent stations because we are interested in finding the areas that are not covered by the existing network, so redundant coverage is not a consideration. By subtracting out the covered areas from the map of the state, we can create a map that shows the areas that are greater than 70 km from any existing station which we can readily see by looking at the unhatched areas in figure 66.

We can perform a spatial analysis to determine the distances to the nearest covered area for those areas that are outside the 70 km range. Fig 67 shows classes of distance from covered areas. Green areas are within 0-70 km of a covered area (i.e. within 140 km of a station). Yellow areas are within 70-140 km of a covered area, and the red are within 140-210 km of a covered area (within 280 km of a station). Thus, putting new stations at the junction between the yellow and green would extend 70 km spacing. Reclassifying these ranges of distances into 3 classes based on their distance from the nearest existing CORS, we can obtain the statistics for this dataset. There are about 191,407 square kilometers without coverage out of the total 380,532 sq. km area of the state. Thus, about half the state has no CORS coverage.

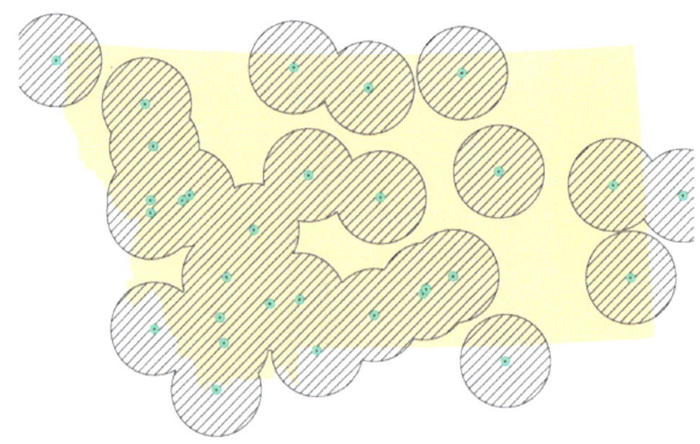

FIGURE 67 70 KM SPACING OF EXISTING STATIONS

To determine the optimum locations to place new stations in order to maximize coverage, we apply our criteria as listed above. We can represent each rule as a dataset, and we can calculate the spatial intersection of these rules. The results will be the best locations for new stations. When we apply the rules we can also give them weights so that the more

FIGURE 66 AREAS WITHOUT CORS COVERAGE, COLORED BY DISTANCE FROM COVERED AREAS (GREEN <= 70 KM < YELLOW <= 140 KM < RED <= 210 KM)

important conditions get higher priority than conditions that may be acceptable though less important. Certainly placing the new stations in areas with greater need would have higher priority than areas of less need, and in most cases, the need is greatest where there are the most people. So population density is an important consideration. High population density will have a higher priority than areas with few or no people. The further away from populated areas then the lower the priority will be. Also, because physical access is important for maintenance, of a station, we prioritize areas that are near roads. In this case, road access is a go or no go situation. If a location is not within 100 meters of a road we will remove it from consideration. Legal access is a consideration, and we can typically obtain access on public lands, if can't find a location on public lands we might be able to negotiate a right of access on private lands. So, in our analysis we assign a higher priority to public lands than private lands, but private lands are still a possibility.

In order to maximize coverage, we want the new stations to be somewhat centered in locations that are within 70 km of existing coverage. Therefore we prioritize areas that are as near 70 km from the edges of existing coverage as possible. We do this with a cost analysis, whereby areas of no coverage that are close to the edge of existing coverage rates a higher cost – thus the further from the edges the better, up to a distance of 70 km.

23.2 TERRAIN

In order to locate the station on level ground that is clear of obstructions by hills or mountains we locate we restrict sites to areas where the terrain is less than 15% slope. We can select such areas by performing a slope analysis on an existing digital terrain model that calculates the percent of slope. After that calculation is completed, we can eliminate those areas that exceed 15%. These areas get a high priority, while areas where the slope exceeds 15% are ignored.

23.3 LEGAL CONSIDERATIONS

Since public funds are being used for this project, we want to ensure that the station is placed on public lands where legal access is assured. For this we query the cadastral database to locate public lands (figure 68). The cadastral database does include fields for ownership classifications (public/private, etc.) as well as subclasses for types of public ownership such as local government, state, government, federal etc. However, for this example we will not

FIGURE 68 PUBLIC LANDS – SHOWN IN LIGHT GREEN.

further reduce this criterion to that level. We will assume that any public lands will suffice. The map of public lands in the areas of no coverage shows that some towns do *not* have public lands.

23.4 PRIORITIZE BY NEED

The map in figure 69 shows the towns in the areas that have no CORS coverage. The large dots represent the towns with population greater than 1,000 people. The smaller dots are the other towns without coverage. The numbers in this map are the population of the town (2000 census).

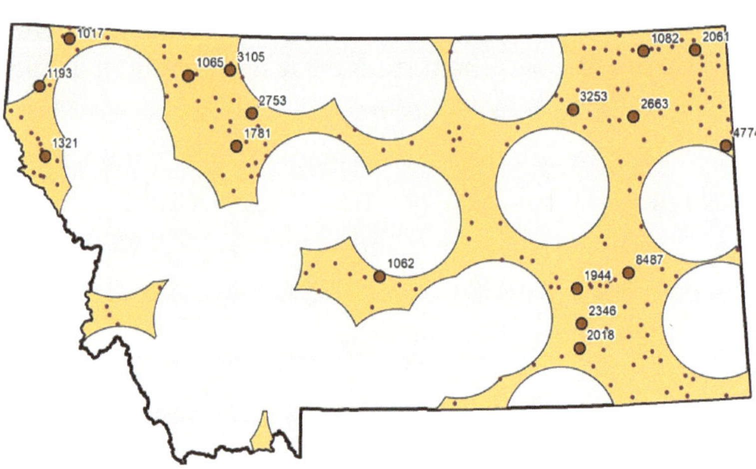

FIGURE 69 TOWNS WITHOUT CORS COVERAGE.

23.5 BRINGING IT ALL TOGETHER

Each of the criteria produces a set of areas that may be appropriate for siting our new CORS stations. We will combine all these criteria to determine the optimum locations for the limited number of sites that we plan. We use the ArcGIS model builder to generate each set of criteria, and to combine them all for a meta-analysis that will calculate the best sites based on all the considerations. The analysis model is shown in figure 70.

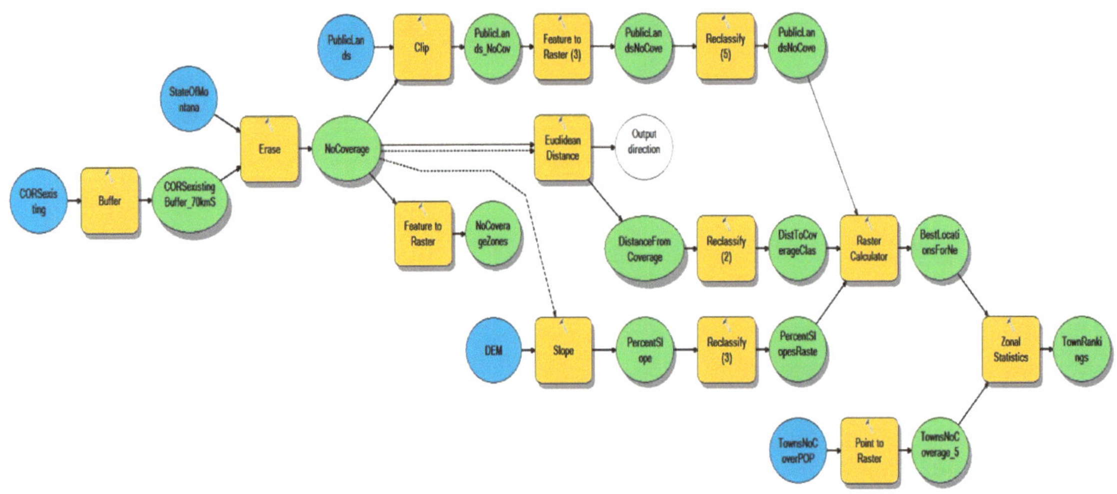

FIGURE 70 DATA PROCESSING MODEL FOR OPTIMIZING LOCATION OF NEW CORS

CHAPTER 24 PARCEL MAPPING

WHAT IS THE PARCEL OR CADASTRAL LAYER?

The parcel layer is cadastral data that is the key to land ownership, parcel size, configuration, land use, improvement values, and other related information contained in federal, state, local government, or public and private agencies. For the purposes of this discussion, a parcel is defined as a contiguous single ownership interest (which may consist of more than one owner) in a legally described real property such as shown in figure 71. A parcel can be comprised of one or more lots or portions of lots in a subdivision, an entire section (square mile of land) in the public lands survey systems, or a single closed figure with a metes and bounds description. Some examples are, a lot in a subdivision, two adjacent lots owned by the same person, an aliquot parts property such as, "all of section 23 in township..." There are other definitions of a parcel but this is the most common usage in GIS. The reason for the "single and contiguous" requirement is a function of the GIS data structure. The important thing to remember about the GIS parcel layer is that it is used in GIS in ways that are similar to how it has been (and still is) used in the hand-drafted and hand-drawn methods.

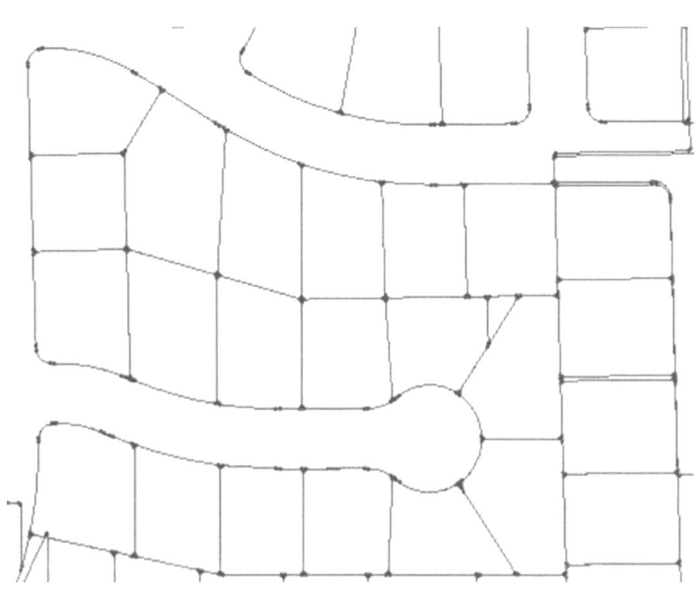

FIGURE 71 PARCEL POLYGONS

GIS differs from hand-drafted maps in various ways - GIS is easier and faster at performing searches and queries (for example, finding who owns a property, or locating vacant lands, ascertaining land values, etc.), it is easier and faster to maintain and edit, and most importantly GIS facilitates the integration of the parcel data with other layers, such as soil type, slope, and infrastructure.

WHO USES PARCEL DATA AND HOW IS IT USED?

The list of users and uses of parcel data is limitless, but typical users are government agencies with land administrative responsibilities, private companies, title companies, real estate brokers, land surveyors, corporate land managers, utility companies, energy exploration companies, and the general public. Some examples of uses are: identifying who owns land along a highway corridor for a road widening project, performing property appraisals, or identifying all the conservation easements in the state.

Parcel data can give information on land uses such as agricultural, residential, park, industrial, or retail. Such information can help identify areas with a potential for development or protection. The GIS parcel layer can be used to identify landowners along a potential route, trail, or automatically generate a mailing list for notification or form letters. Parcels are used for rural addressing to help identify where improvements exist which may need to be addressed, as well as who the owner is so that they can be notified of their new address. The ownership information can also be used to develop a list of residents to be evacuated in emergency situations, such as the wildfires in the American West this past summer.

ATTRIBUTES OF PARCELS

Parcel data will include a variety of tabular information connected to the parcel. The type of data varies in each jurisdiction, but listed below are some examples of the types of database fields that are typically associated with a parcel and figure 72 shows a typical GIS map of parcels with attributes.

- property type is a code that best describes the actual present day use
- improvement classification code identifies the classification of taxable property found within a parcel
- the number of living units present in the dwelling described on this card
- local zoning of the parcel
- the land value
- the total fallow acreage of the entire parcel
- the total grazing acreage of the entire parcel
- the total wild hay acreage of the entire parcel
- the total timber acreage of the entire parcel
- the total farmstead acreage of the entire parcel

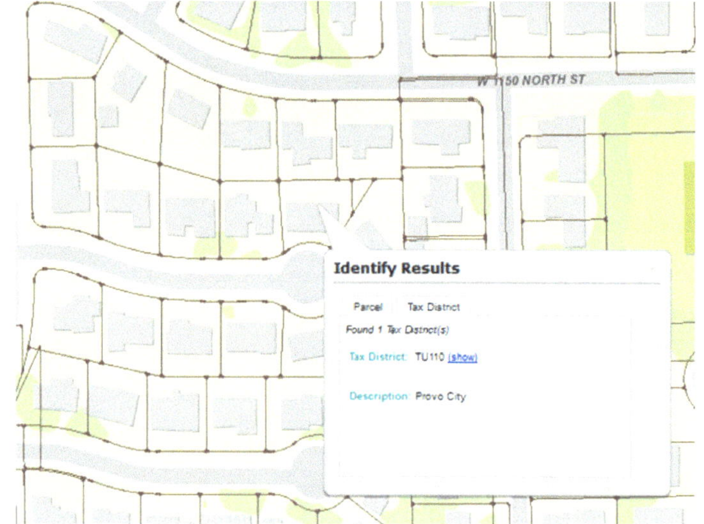

FIGURE 72 PARCEL WITH TAX DISTRICT ATTRIBUTES

- the total irrigated acreage of the entire parcel
- the code that describes the cost per acre to irrigation
- a code that describes the proximity of parking available
- the actual story height of the dwelling
- the code that best describes the type of roof

DIFFERENT USES OF PARCELS

Surveyors' uses of parcel data may differ from other users. A surveyor may be more interested in the history of a parcels' boundary line and have more concern regarding the placement and geometry of a parcel than in its use or value. The surveyor's perspective is, in some ways, a more detail-oriented technical view of the construction and geometry of a parcel. The surveyor wants to know where a particular line originated, and its length and direction. Most other users may be less concerned about the specific geometry and more concerned about the attribute characteristics of the parcel. Most users are more interested in the database information associated with the parcel than they are in the parcel's geometry.

Figure 73 shows a sample of a parcel record of the Republic of Nigeria Mining Cadaster Office. This example is a mining lease interest in a quarry. The fields include Type, Status, Name, Grant Date, and others. A GIS query could be performed on any of the data fields listed, such as "show leases due to expire within the next year", or "show valid leases smaller than 5 sq. km".

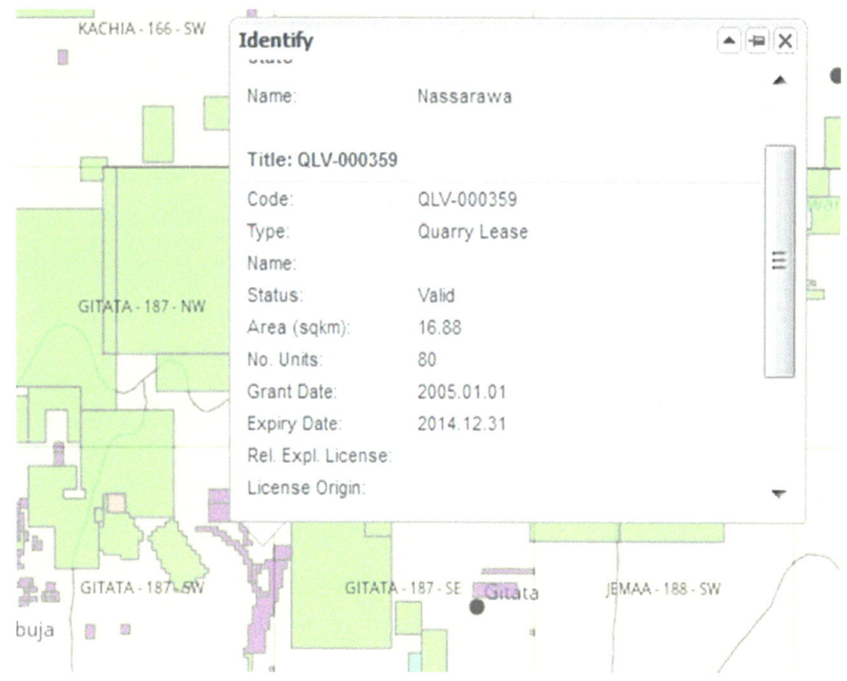

FIGURE 73 IDENTIFYING A PARCEL

Queries may be based on the tabular data as in the example above, or based on location, or any combination of tabular data and spatial relationships. For example, "show parcels that are within 100 feet of a selected road segment" as shown in figure 74.

Parcel layers contain far more than just location and geometry information, which is why the cadastral layer is so important. GIS can generate maps based on any of the database-coded information. Depending on how the parcel layer was developed, it might not directly provide accurate location and geometry data such as: the exact location (coordinates) of the property corners, the bearing and direction of parcel lines, or acreages. That is, the parcel lines themselves might not be constructed using the historic and/or legal geometry. In these cases, the parcel configuration is a general representation of the legal geometry, in fact more of an index to the tabular information. This does not diminish the parcel layer's usefulness for the previously described purposes, but it may not support some surveyors' purposes.

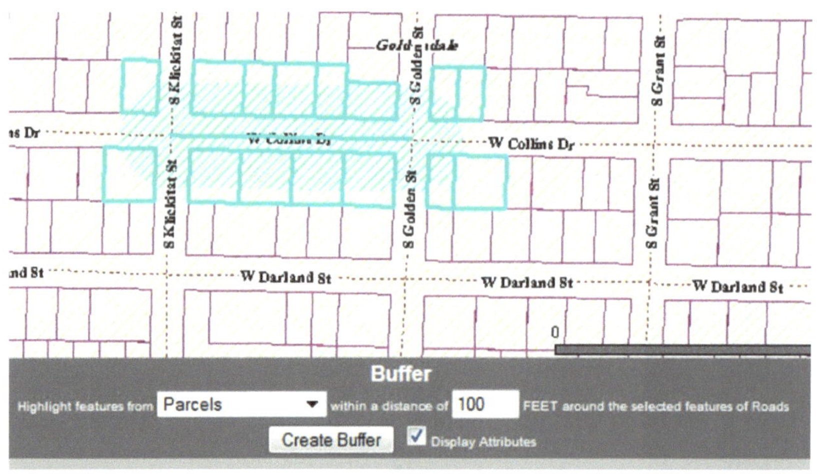

FIGURE 74 SPATIAL QUERY – PARCELS WITHIN DISTANCE OF SELECTED ROAD SEGMENT

DIFFERENT USES

Not all parcels are created the same way. A major difference between a surveyor's perspective of a parcel and GIS is the use of projections and coordinate systems in GIS. Surveyors typically work in small areas when dealing with parcels and thus work in plane (or flat geometry) surveying. GIS mapping, however, will typically cover much larger areas such as a city, a county, a national forest, or an ecosystem that may include several drainage basins. When larger areas are the extent of interest then a mapping projection may be implemented in order to conform the geometry of a non-flat world into geographic space. Moving a parcel into a projection typically introduces geometric changes in bearing rotations and geometry scaling, therefore altering the parcel geometry. This does not mean, however, that the GIS cannot provide the legal and historic parcel geometry information. The legal bearing and distance, as described by the plat of survey or deed, can be an attribute of the line work or the parcel. Alternately (or additionally), the GIS can have a hot-link from the parcel to the scanned image of the plat or survey or deed that created the parcel in order to present the legal and or historical information. Either of these methods allows the user to obtain the legally recorded information on parcel geometry or size by simply clicking on the GIS parcel map.

SURVEYOR ISSUES AND CONCERNS

Other concerns that surveyors have about parcels tend to revolve around regulatory implications of GIS mapping. Zoning, floodways, wildlife protection, and other constraints may be placed on a parcel due to the parcel location. Usually these types of determinations are completed in a GIS by overlaying various administrative layers upon the parcel layer. For example, the Federal Emergency Management Agency (FEMA) flood data which shows the 100-year and 500-year flood may be overlaid to show whether or not a particular property is within the regulatory floodway (figure 75).

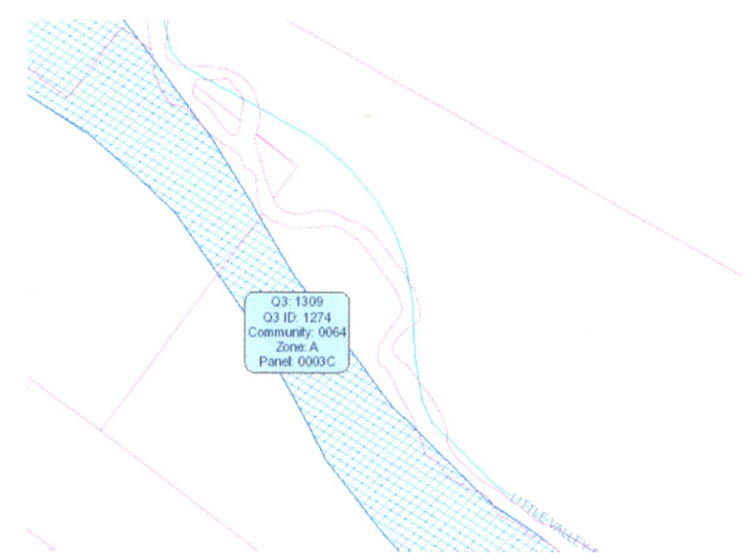

FIGURE 75 FLOODPLAIN ADMINISTRATIVE MAP (HATCHED AREA) OVERLAIN ON PARCELS (PINK LINES)

Surveyors are concerned that the GIS layers may not be spatially accurate enough to reasonably determine such locations, particularly when those determinations will have a financial impact on the property. Questions may arise such as, *ARE THE PARCEL LINES REALLY IN THE RIGHT PLACE AND IN CORRECT CONFIGURATION*? or *ARE THE FLOODWAYS CORRECTLY MAPPED*? These concerns are equally valid for hand-drafted maps as well, but as stated earlier, GIS facilitates these types of overlays, which highlights the differences between datasets. The good news is that GIS makes these spatial comparisons much easier than the compilation of hand-drafted maps. After all, these comparisons were being completed long before GIS came along, and will continue to be performed with or without the benefit of GIS. One of the best remedies to address everyone's concerns about spatial accuracy (whether absolute accuracy or relative

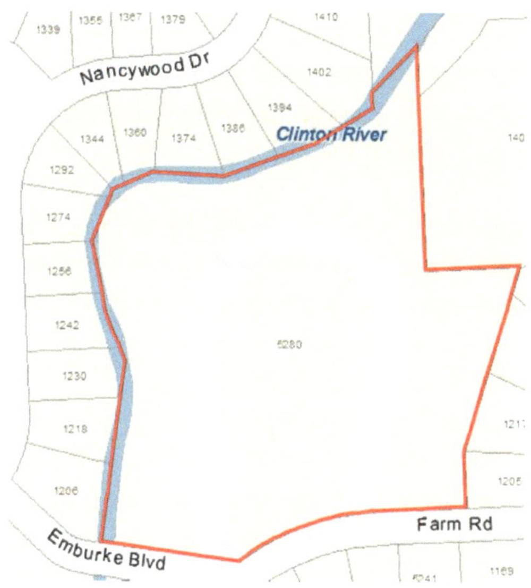

FIGURE 76 LARGE PARCEL (RED OUTLINE) WITH RIVER AND ROAD BOUNDARIES.

accuracy) is for more surveyors to be directly involved with building and maintaining parcels.

GIS TECHNOLOGY STANDARDS AND ISSUES

GIS technology models the real world by using graphic representations that describe things as points, lines, or polygons. Parcel boundaries represent areas which in GIS are most usefully described by polygons. The way GIS understands what is in and what is out of a polygon dictates that the polygon must be closed geometrically.

Figure 76 shows a parcel (red outline). This particular lot is bounded by roads, a river, and other parcels. The parcel geometry must be a closed polygon area in order to be a proper GIS polygon and will typically contain at least one attribute - a unique identifier. Also, the final GIS parcel layer must be a seamless mosaic of the individual parcels. That is, there must not be any gaps or overlaps of adjacent parcels. Gaps and overlaps be challenging to resolve when mapping parcel data that come from different sources. This is especially true when documents use different points of beginning or basis of bearings to locate and orient the parcels. But these issues *must* be resolved for a GIS, in order to achieve the goal of a seamless mosaic of ownership. Everyone who has tried to create a parcel layer knows that some record geometry will be distorted by this process. Other issues that may also change parcel geometry include converting map data to other coordinate systems (for example, state plane) or projections (UTM), or simply fitting the map data into survey or mapping control.

SOURCES OF INFORMATION

The first place to start for the parcel information is assessor maps if available. Assessor maps will show ownership boundaries and may also show easements, rights of way; hydrography, map control points, subdivisions, and possibly include annotation of parcel geometry, lot numbers, surveys, and other useful references. The assessor map historically has been a hand-drafted compilation of the record data such as deeds, surveys and plats. The original map complication may have used aerial photography as the source for supplemental information (roads, buildings, rivers, etc.). Where assessor maps are not available, or if they are incomplete or in conflict, then one must refer to the record data (that is, the original and legal source for the parcel locations and configurations) such as deeds and surveys in order to research the location and geometry of parcels. If at all possible, these resources should be scanned to facilitate research, access, retrieval and use. It is far quicker and easier to pull up a digital deed then to have to dig through a pile of papers to locate a legal description. If scanning is not an option, then paper copies of all relevant documents should be obtained. These documents may be stored at the county courthouse, the county surveyor's office, land records office, or assessor's office or other locations. Typical data sources required for parcel mapping are listed below:

- Deeds
- Subdivision plats
- Court orders
- Assessor maps
- Road records
- Surveys

MAPPING CONTROL

Mapping or survey control is essential to the cadastral layer. The control is the framework upon which all the parcels are referenced. Typical controls for parcel mapping in the USA are listed below. The conversion process may use one or more of these sources of control:

- Public Land Survey System (GCDB– Geographic Coordinate DataBase, or USGS topographic maps)
- National Geodetic Survey
- USGS Quad maps (PLSS corners, fence lines, occupation lines, etc.)
- Local control (city, county, private)
- Orthophotos, planimetrics (road centerlines, road edges, city block boundaries, hydrography, etc.)
- GPS road centerline
- Subdivision or lot corners

The order in which the control is introduced to the data conversion process varies from one agency to the next. Some start with control and build inside that framework. Others build each map individually then place them into the control framework. There are many variations of these two basic methods, but typically, a least squares type of adjustment is performed somewhere during the conversion process any error residuals are then examined for acceptance. It is not unusual to discover large, unanticipated errors somewhere in the project area when going through the conversion process, which may necessitate acquiring supplemental control to resolve.

METHODS FOR CONVERSION

The conversion strategies vary widely, but fall into two basic groups: digitizing the source documents, or data entry of the record information (COGO or COordinate GeOmetry). COGO is more accurate representation of the geometry as whown in figure 77.

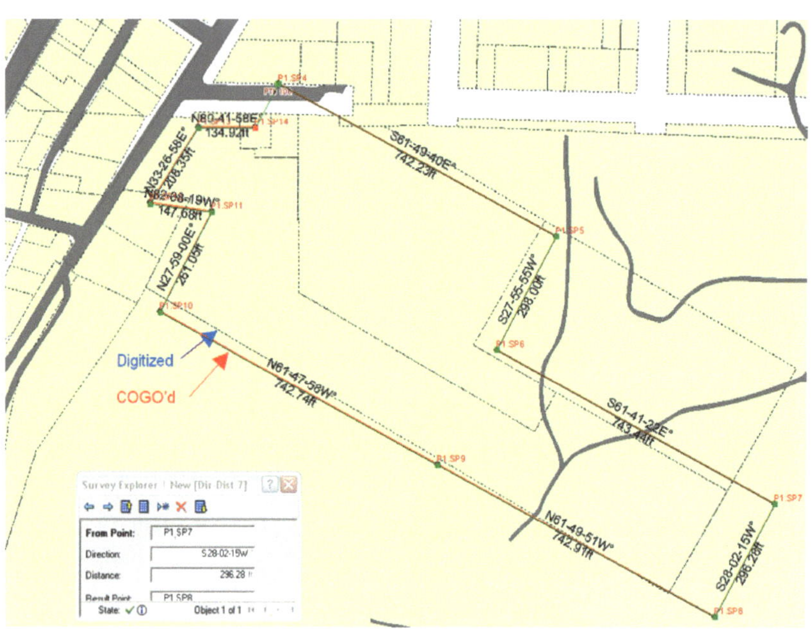

FIGURE 77 DIGITIZING VS. COORDINATE GEOMETRIC METHOD FOR MAPPING

Digitizing may be performed on the hard copy maps (Mylar or paper) taped to a digitizing board, or, more preferably, the source data can be scanned and used as a backdrop for on-screen digitizing. There are also automated line-following algorithms that facilitate batch processing of the digitizing work.

The COGO work may be performed with COGO software that interfaces with a software such as AutoCAD or ArcInfo, or the data may be either entered into a spreadsheet or database which can later be imported into a GIS, or with geometry construction tools in either computer aided drafting (such as MicroStation or AutoCAD) or GIS software. The scanned images of the source data can also be used as a backdrop for the COGO data entry in order to speed up reading the record information, or in the case of plats or assessor maps, to visually track both the level of completion and as a quality check on the lengths and bearings of the line work. Additionally, there are intelligent systems that can "read" and plot legal descriptions from scanned documents (after OCR–optical character recognition–processing), or from other text files.

CAD TO GIS

Parcel data that are created in a CAD format such as AutoCAD's DWG or MicroStation's DGN can be converted into GIS formats. The difference between CAD and GIS is that the GIS format has the ability to understand spatial relationships among features. It can indicate, for example, the number of fire hydrants within 300 feet of another fire hydrant, or the number of wells that are on parcels of land not designated for agricultural use.

Some of the considerations, when converting from CAD to GIS, are that a single layer or level in CAD may contain different types of objects (points, lines, polylines, annotations), which GIS will characterize differently. These different types of objects will typically be segregated out when converted to GIS format. Therefore, parcel polylines in CAD will be converted to GIS polygons and/or lines. The text (in the CAD file) may become annotation in the GIS format. Suffice it to say that the conversion is rarely a seamless one, and usually requires some programming or customization to achieve the desired end result.

TYPES OF BOUNDARIES

Parcel boundaries may be comprised of metes and bounds, lot and block of a platted subdivision, land grant, or aliquot parts of the US Public Lands Survey System. Metes and bounds descriptions may include bounds based on riparian features such as rivers, creeks, or lakes, or may have public rights of ways as bounds (e.g. *to the easterly right-of-way of county road…),* or topographical calls such as to ridge tops, etc. Some of these boundary types can be problematic to use when creating parcel from them. Riparian boundaries, for example, may be mapped in GIS format at and readily available but the scale for which they were mapped may not match the parcel mapping. Riparian features are often available in national spatial data sets but these are usually created using small scale mapping which may not suffice for large scale parcel mapping. The difference in mapping scale can create issues with misalignment and position of riparian boundaries for parcels. It may be necessary to perform new mapping to match the scale and spatial accuracy requirements of the parcel mapping.

COMPLETING THE GIS DETAILS

As mentioned earlier, one of the great advantages of GIS is the ability to assign attribute information to graphic features, in order to be

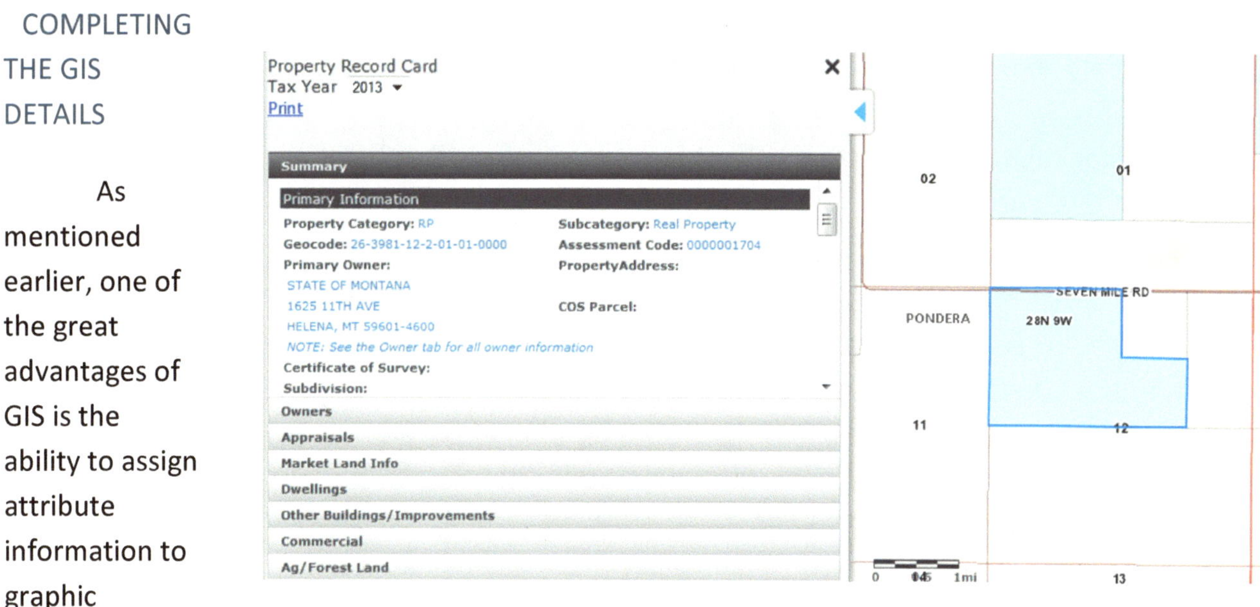

FIGURE 78 ATTRIBUTES OF OWNERSHIP

able to perform query and analysis of the data. To accomplish this, each parcel polygon must be assigned a unique identifier (for example, a parcel ID number or assessor number). This can be done with text in a CAD file or performed as a data entry task in the GIS attribute table. Once that work is done, the parcel layer's attribute table can be related to any other table that shares those same parcel IDs. Usually the parcels are related to a table of ownership and valuation information but any table may be used (a spread sheet of client information for instance). Depending on the software used and its data model, more than one table could be joined simultaneously. For instance, a parcel layer may be related to an ownership table from the assessor's office as well as a table of land management from the fish and wildlife office. The parcels can also be assigned hot links that will pop up image data (such as a picture of a house or copy of the deed). All these tables are related to the individual parcels via the parcel identifier.

MAINTAINING AND IMPROVING PARCEL DATA

It is common knowledge that parcel configurations are not static. Properties are subdivided, common boundaries are altered, lot lines shift. The parcel layer is dynamic and must be maintained to keep it up to date. The accuracy of existing parcel geometry can also be improved by introducing new and better survey and mapping control, then adjusting the line work. Surveyor's information greatly benefits the update, maintenance and improvement of parcel data.

Important Points to Remember

- GIS needs closed polygons: There can be no gaps or overlaps.
- Concerning record data vs. cartographic representation: GIS is a graphic representation of the source data and does not purport to be the legal definition of parcel location, configuration or history.
- Concerning rubber-sheeting (that is, least squares adjustment): It is not uncommon to alter some of the source geometry in order the meet the requirements of a seamless, consistent database. This is OK. There are mechanisms for maintaining both the source reference information, as well as the record geometry used to construct each line and parcel.
- When relating parcel ID to a database: The parcel layer blossoms from a dumb graphic to a true GIS layer when each parcel is assigned a unique identifier that can act as a key to records in other tables.

CHAPTER 25 USING GIS TO MANAGE A PROJECT

Because GIS is a good tool for visualizing, analyzing, managing, and communicating information, it is helpful for planning a development project and for managing the progress of the various phases of the project. Additionally, GIS can bring information together from a variety of sources to help one in site selection based on demographic data, infrastructure support, potential markets, etc. Some governments provide site selection mapping applications free online to support economic development (see figure 79).

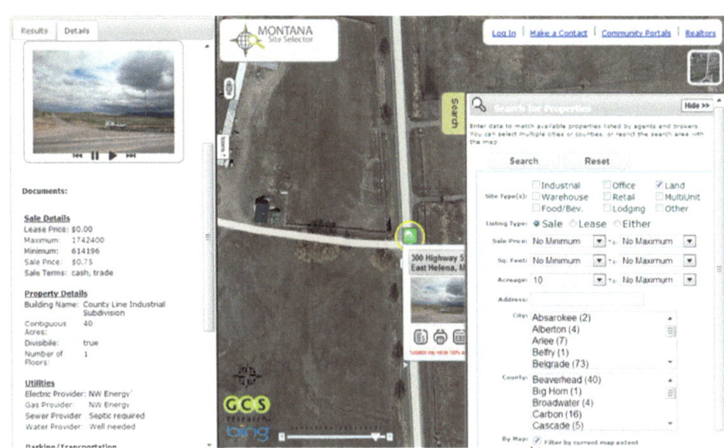

FIGURE 79 SITE SELECTOR INTERACTIVE GIS WEB SITE

Research phase – discover which surveys and plats have been done in the area, what survey control exists.

In the conceptual phase of a project GIS can help one understand where a project is located and the general lay of the lay of land. GIS can help one to understand the physical and administrative conditions that exist on a property prior to drafting a plan such as zoning and easement constraints. GIS analysis provides tools to further understand existing conditions, such as calculating the degree of slope of the terrain, the direction of flow of surface and ground water, calculating distances to existing infrastructure for roads, water, sewage, and electrical services.

GIS helps the planning phase by analyzing the optimal location for roads and infrastructure services, estimating earth volumes, Analyzing the potential market for the developed houses, or retail or commercial structures, for planning various features of a development.

GIS can communicate what the plan is, its impacts on the neighborhood, and the schedule for project's phased implementation. The communication can occur through various media such as printed maps, electronic maps such as geo-registered PDF and graphic maps which may be emailed or posted on the web as static maps, or as interactive maps which can be dynamically updated which is a useful way to track and communicate the progress of a project's phases. Here we look at some examples of how GIS helps plan and manage a civil engineering development project.

25.1 UNDERSTANDING THE EXISTING CONDITIONS

GIS can bring together existing information in a way that helps one to understand the conditions and status of the property as represented in figure 80. Typical themes include the physical characteristics of the property, the administrative constraints, demographic information which may influence the type of development, and the existing built environment including existing infrastructure to which one may need to connect new services. Some of these themes are listed below. Nearly all of these may already exist in digital form and possibly be available for download or by connecting to map services. Others may be created via GIS processing.

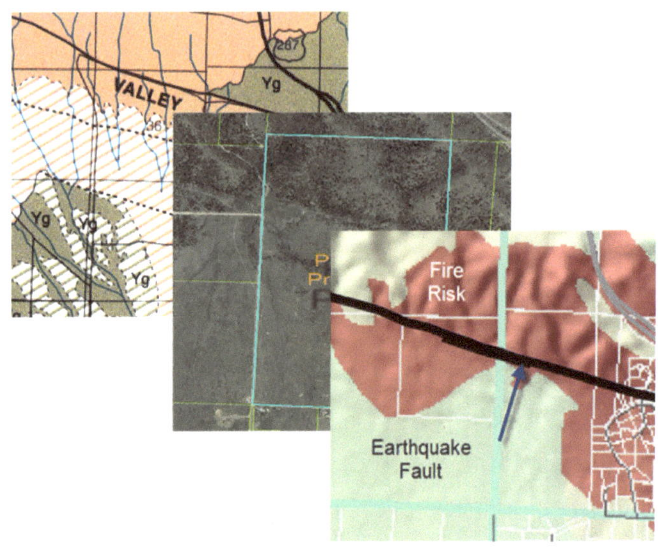

FIGURE 80 THEMATIC MAPS OF EXISTING CONDITIONS: GEOLOGY/ZONING/HAZARDS

PHYSICAL

- Property location
- Soil & Geology
- Slope*
- Aspect*
- Existing vegetation
- Hazards (flood, earthquake faults etc.)
- Climate (precipitation, hours of sunshine, number of freezing days)
- Hydrology* (where water goes)
- Hydrography (where water is)

BUILT ENVIRONMENT

- Existing Roadways
- Existing Water and Sewer
- Power, Internet

ADMINISTRATIVE

- Land Ownership
- Existing Plats and Surveys
- Public vs. Private Ownership
- Road rights of way
- Zoning

DEMOGRAPHICS

- Population numbers and location
- Population Income
- Average home price
- Employment Rate

25.2 ESTIMATE A PROJECT

Prior to beginning a project it is essential to estimate the cost to provide the necessary services. GIS can help develop cost estimates in a number of ways but is particularly useful for factors based on geography, location and distance.

PROJECT LOCATION

Location of a project with respect to the location of the office where services are based is an important factor in project cost. Things to consider are the project distance from the office; the cost to travel to and from the project area, the number of miles and time it takes to go to and from the project, and the number of trips. Projects that are a considerable distance from the office may require staying overnight for a short while or extended periods of time. Thus mileage and per diem cost factor into the estimate as well as travel time.

Project travel time and distances can be readily estimated using any of the free online mapping and directions tools such as Bing Maps, Google Maps, MapQuest, Yahoo Maps, etc. These applications can provide mileage distances and estimated time which one can then apply cost factors to for mileage costs, and crew travel time to and from the project area. These applications can also provide information on lodging in the project area if necessary, including estimated lodging costs.

FIGURE 81 PROJECT AREA SURVEYS & PLATS

PROJECT AREA CADASTER

Existing surveys, plats, deeds, zoning, easements, rights-of-way and other administrative impacts on the project area can be easily manage, visualized and integrated using GIS like in figure 81. GIS is particularly helpful for this because GIS was designed to integrate various data for planning purposes. Many of these data, such as rights of way, zoning, conservation easements, etc. may already exist in GIS form. Those that are not in GIS form may exist in CAD format or graphic format, which with proper geo-registration can be displayed in GIS and thus overlain with other data.

PROJECT AREA TOPOGRAPHY

Online map applications and mapping applications such as Google Earth can provide can provide perspectives on the approximate topography of the project area. The terrain displayed in these applications is usually no better than 10 meter to 30 meter resolution, thus these are insufficient for design, yet they can provide a general idea of how flat or steep a project area may be which is helpful in developing estimates for providing services. Existing aerial photography can also depict whether the project area is timbered, has existing buildings on it or is clear and open, which has an impact on the cost to develop a property. While these factors may not be directly quantifiable, they can be taken into account when developing a project plan of work and estimate. For instance, if it's necessary to clear trees and brush, one can estimate the linear distances or area extents of the areas that must be cleared and then estimate the amount of time necessary to perform that work.

Additionally, water bodies and streams can affect the scope and type of layout as well as affect the cost of getting around on a job site. Water bodies and streams can be seen in freely available aerial photography and in some satellite imagery.

THESE INFORMATION CAN BE INTEGRATED

All the characteristics affecting the project property can be brought together in GIS for the purpose of understanding the various elements individually and combined in various ways to obtain a comprehensive picture of their impacts on project cost, and then used to estimate the cost of doing the project. The ability to integrate geographic data sets that represent a variety of thematic concerns is the fundamental power of GIS because it aids in human understanding of the spatial relationships among disparate data that may be related only by their sharing a common landscape.

25.3 ANALYZE

The real power of GIS is its ability to analyze various things which may be essential to understanding the existing conditions. Some examples of GIS analysis tools are listed here that are representative of what one may do using GIS.

SLOPE ANALYSIS

One can estimate the amount and direction of slope based on elevation data such as contained in a digital elevation model, or digital terrain model or Lidar data. Existing digital elevation data are available for most of the earth from a variety of sources, such as the United States Geological Survey (http://nationalmap.gov/viewer.html) which varies from 10 meter to 90 meters, or the Virtual Terrain Project (http://vterrain.org/). These data are sufficient for small scale mapping. Large scale data may be available locally.

Data input -> generate slope, generate aspect -> display -> get statistics

BUFFERS

GIS can quickly generate buffer areas - offset distances around a feature such as shown in figure 82. Streams and water bodies can be buffered as may be other linear or polygonal features to determine offset distances. For example a road centerline can be buffered with a right-of-way width, or a stream buffered for and environmental impact constraint. The resultant buffered areas can be used to enforce exclusion restrictions, can be used calculate the amount of affected areas (acres/hectares), used to calculate volumes of earth or timber removal.

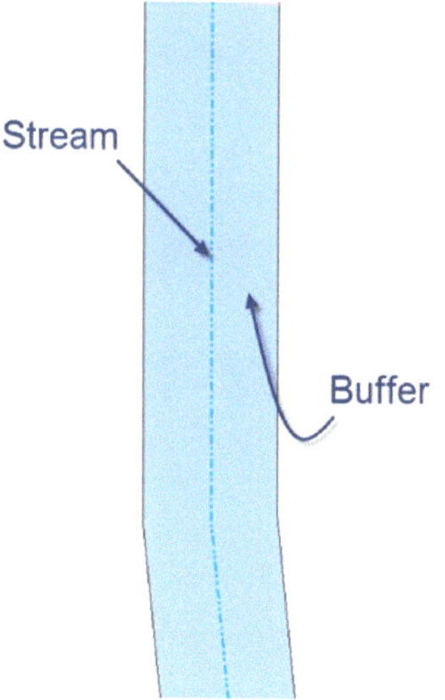

FIGURE 82 BUFFERING A STREAM

25.4 PROJECT PROGRESS TRACKING

The progress of a multi-phase survey or development project can be mapped to show the status and location of the work to do and the percent of completion of various tasks. Project progress can be documented in a spreadsheet or database which can be visualized on a map. An example of this is the Texas Department of Transportation Project website that communicates the phases and progress of highway projects using a table approach as shown below.

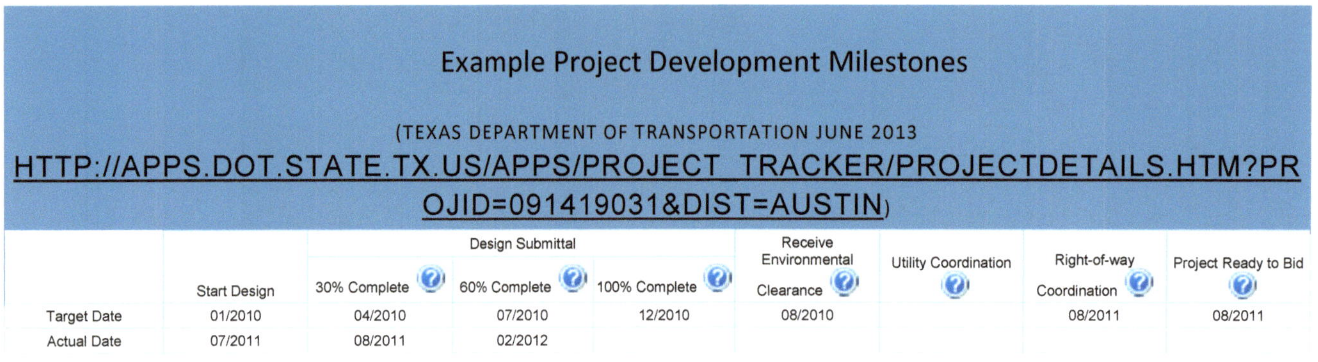

Example Project Development Milestones

(TEXAS DEPARTMENT OF TRANSPORTATION JUNE 2013 HTTP://APPS.DOT.STATE.TX.US/APPS/PROJECT_TRACKER/PROJECTDETAILS.HTM?PROJID=091419031&DIST=AUSTIN)

		Design Submittal			Receive Environmental		Right-of-way	
	Start Design	30% Complete	60% Complete	100% Complete	Clearance	Utility Coordination	Coordination	Project Ready to Bid
Target Date	01/2010	04/2010	07/2010	12/2010	08/2010		08/2011	08/2011
Actual Date	07/2011	08/2011	02/2012					

25.5 COMMUNICATE DEVELOPMENT PLAN

Communication is an important component of every project. Internal communication among teams within a company or agency, inter-department communication, communication with government officials, and with the public regarding what will happen, where it will occur, and when must be done in timely and effective manner. GIS is a great help with communicating the right information to the right parties and when used properly everyone can be kept up to date on status, and scheduling for smooth operations of large projects.

STATIC MAPS

Static maps such as the one shown in figure 83, are easy to produce and can readily communicate important aspects of a project.

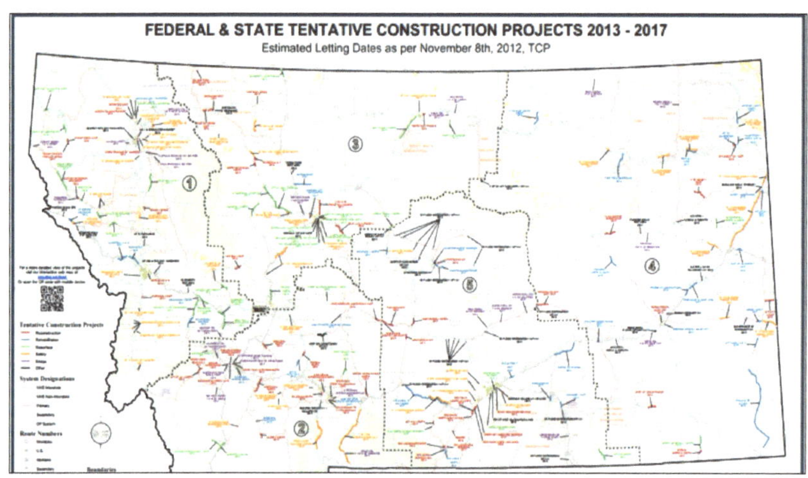

FIGURE 83 STATIC PROJECTS MAP (PDF)

The static maps can be

produced and published as Adobe PDF, or graphic formats such as JPEG and GIFs. These static maps can be produced periodically throughout a project at regular intervals, such as weekly progress reports or daily plans for where work crews will be going and what they plan to accomplish, or they can be produced to communicate changes of status for major project phases.

Static maps may also include a map series which may show the same themes for different locations or areas, or various thematic characteristics for the same area. Map series can be generated a few ways in GIS, and some instances the data can programmatically drive the content and map extents of each map within a series.

INTERACTIVE MAPS

Interactive web maps like the one shown in figure 84, are useful when there are larger areas with multiple projects, or when more detailed information must be communicated to a disparate audience that has differing interests. Interactive maps typically provide and interface that allow the use to select their geographic area of interest and to drill down to the types of projects or aspects of the project in which they are most interested.

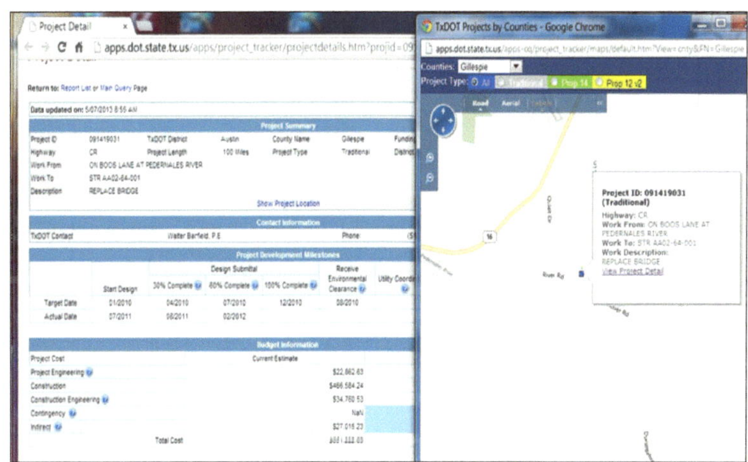

FIGURE 84 INTERACTIVE WEB MAP

3D VISUALIZATION

Three dimensional (3D) visualization of the development plan like the one shown in figure 85 which shows buildings and water wells, can be very help in understanding the scope and impact of the development concepts of the plan. 3D visualization provides a good perspective of the location and extent and impacts of the various components of the development plan, such as cuts and fills of a roadway location, the location of a pipeline through the property, the number of buildings and where they may be constructed, and the impacts on view shed.

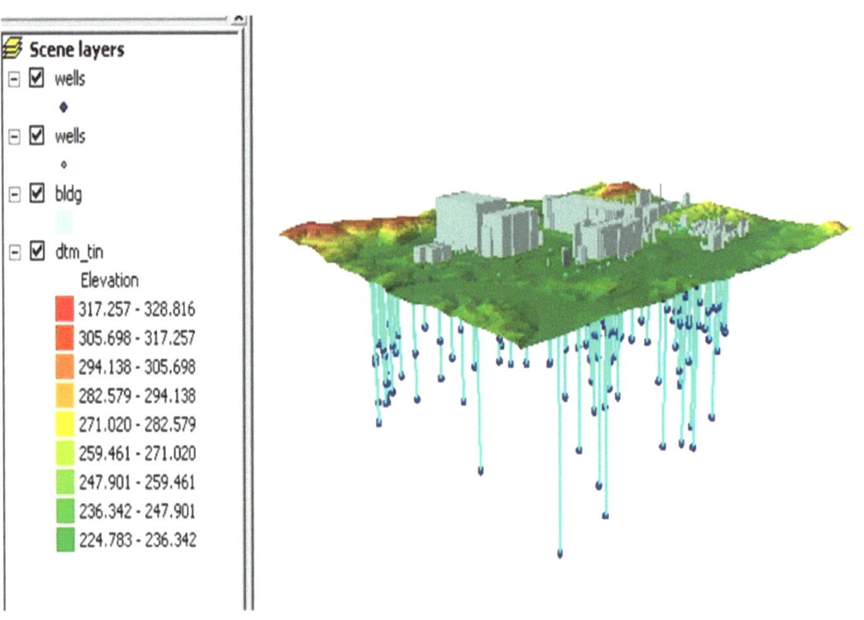

FIGURE 85 3D VISUALIZATION OF PROPOSED BUILDINGS AND WELL DEPTHS

CHAPTER 26 USE GIS TO CREATE FORM LETTERS

One very simple application that can save a huge amount of time is to use GIS to select a list of information to use in form letters. For example, to provide property owners of an upcoming survey in their neighborhood, a surveyor could use GIS as shown in figure 86 to select the adjoining properties, obtain the land owner names and mailing addresses, and merge that information into a form letter to automatically print letters and envelopes. This saves time in the following ways,

- Access to GIS property information saves the time required to travel to the courthouse and search through the record, looking up each property on a hard copy map, then searching through the records for the owner name and mailing address.
- The ability to export a table of selected records from GIS saves the time otherwise needed to transcribe the owner names and mailing addresses by hand.
- The ability to import the data from GIS into a word processing document for a form letter saves the time of manually typing in each owner's name and address.
- Because the records are pulled directly from the database into the word processors, typographical errors are avoided.
- If a record or two is missed, it takes only a few minutes to go back to the GIS and select them, which is much more expedient than having to make another trip to the courthouse.

FIGURE 86 CLIENT PROPERTY TO SURVEY SHOWN IN YELLOW, ADJOINING PROPERTIES SHOW IN ORANGE.

Of course in order to use this kind of data, the surveyor must have access to the GIS information, and the GIS software that provides a means to select records and export the selected records to a file. Presently there are at least 3 ways to get this information. The first is to obtain a copy of the data in GIS format from the state or local agency. The second way is to connect to the data as a map services provided by the data hosting agency. The third way is access the data via web mapping services that provides a means to download selected records.

The important components are the ability to query the GIS layer, such as by owner name, property identifier, road name, or a geographic search (e.g. a map selection of a property and the adjoiners, or by distance from the property) and the ability to save or export the tabular data of the selected records to a local file.

The basic steps are to use the GIS to select the records of interest by using a spatial selection, then export the tabular data for those records to a table (such as a dBase format file or delimited text file), use a word processing software to create a form letter using the field names of the table, merge the data into the form letter, then print the letters. This same process also applies to printing envelope forms for addressing the mail.

What follows is an example of how to do these steps.

STEP ONE - SELECTING THE RECORDS IN GIS

GIS offers many ways to search for and select records. One may query the database on attribute information, such as land owner name or property identifier or perform spatial searches, such finding all parcels that adjoin (boundaries that touch), or are within a certain distance of a selected property or all parcels that are along a certain stretch of road. Figure 86 shows a property selected in GIS. Since this is the property to be surveyed, we will want to inform the adjacent property owners that our crews will be in the area. We use the GIS spatial query function to select those properties that share a boundary with the parcel we are going to survey and shows the properties that share those boundaries.

Figure 87 shows the attributes (i.e. tabular information) of the parcels that were selected. This table has the owner name and mailing

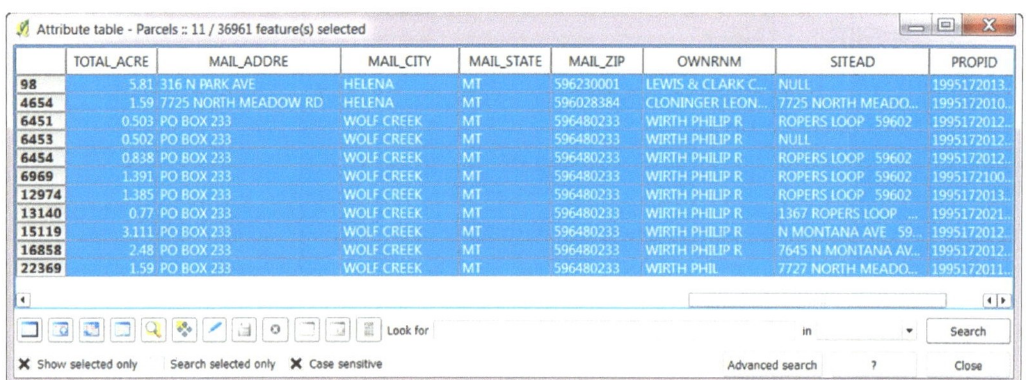

Attribute table - Parcels :: 11 / 36961 feature(s) selected

	TOTAL_ACRE	MAIL_ADDRE	MAIL_CITY	MAIL_STATE	MAIL_ZIP	OWNRNM	SITEAD	PROPID
98	5.81	316 N PARK AVE	HELENA	MT	596230001	LEWIS & CLARK C...	NULL	1995172013...
4654	1.59	7725 NORTH MEADOW RD	HELENA	MT	596028384	CLONINGER LEON...	7725 NORTH MEADO...	1995172010...
6451	0.503	PO BOX 233	WOLF CREEK	MT	596480233	WIRTH PHILIP R	ROPERS LOOP 59602	1995172012...
6453	0.502	PO BOX 233	WOLF CREEK	MT	596480233	WIRTH PHILIP R	NULL	1995172012...
6454	0.838	PO BOX 233	WOLF CREEK	MT	596480233	WIRTH PHILIP R	ROPERS LOOP 59602	1995172012...
6969	1.391	PO BOX 233	WOLF CREEK	MT	596480233	WIRTH PHILIP R	ROPERS LOOP 59602	1995172100...
12974	1.385	PO BOX 233	WOLF CREEK	MT	596480233	WIRTH PHILIP R	ROPERS LOOP 59602	1995172013...
13140	0.77	PO BOX 233	WOLF CREEK	MT	596480233	WIRTH PHILIP R	1367 ROPERS LOOP ...	1995172021...
15119	3.111	PO BOX 233	WOLF CREEK	MT	596480233	WIRTH PHILIP R	N MONTANA AVE 59...	1995172012...
16858	2.48	PO BOX 233	WOLF CREEK	MT	596480233	WIRTH PHILIP R	7645 N MONTANA AV...	1995172012...
22369	1.59	PO BOX 233	WOLF CREEK	MT	596480233	WIRTH PHIL	7727 NORTH MEADO...	1995172011...

Look for in Search
Show selected only Search selected only Case sensitive Advanced search ? Close

FIGURE 87 TABULAR DATA OF SELECTED PROPERTIES.

address, and the location and abbreviated legal description of the property. The GIS as other information about the properties as well, but these are the ones we will use to generate the form letters.

STEP 2 – EXPORT THE SELECTED DATA

Once the parcels are selected, the attributes of those parcels may be exported to a table. In this case the data are exported to an Excel spreadsheet, which will contain the owner and property information we need to create the letter. The file is saved on the computer so that we can later pull the information it into the form letter.

STEP 3 – CREATE THE FORM LETTER

Open a word processor such as MS Word, create a form letter, and then use the mail merge tool to select your exported table as a data source for the form. MS Word allows one to browse for an existing file that contains our ownership and property information. Make sure that the file name is not longer than 8 characters or Word won't load it.

STEP 4 – INSERT THE FIELD NAMES AS PLACE HOLDERS,

Once we have connected to our exported table, we can then insert its fields into the appropriate places in our form letter (figure 88). Our form letter will have place holders for the owner's name, street address, city, state, zip code for mailing purposes, and a short legal description of the property so that owner knows which property we are referring to. Note that the owner's mailing address might be different from the location of the property in question.

Dear **«OWNRNM»**
«MAIL_ADDRE»
«MAIL_CITY», «MAIL_STATE» «MAIL_ZIP»
Regarding your property at **«SITEAD»**

We are writing to inform you of our plans to be in the area to survey the property adjacent to your property at **«SITEAD»**

If you have any questions please contact me.

Thank you.

Your friendly local surveyor

FIGURE 88 FORM LETTERS WITH FIELD PLACEHOLDERS

STEP 5 – MERGE THE DATA TO THE FORM

After the fields

are set up in the form letter we can perform the merge and automatically print (or preview) all the letters (figure 89) with the push of a button. Note, the form letter can be reused, and if the next table exported is saved with the same name as the one previously used in MS Word, then you can skip steps 1 & 4 in the future and just use a stock letter for any future notifications. In that case you would need to use the GIS to create the selection and export of records, and then save that export table to a file of the same name so that when you then open your previously created form letter everything will be all setup and ready to print using the new records. In addition, as mentioned earlier, this method can also be used to create form envelopes for mailing.

The power of GIS is quite evident in how it can make mundane tasks simple, and very quick. The ability to perform a spatial query to discover the relationship between properties in an area, and then to use the information that is in the GIS in a variety of ways, demonstrates why GIS has nearly become a standard desktop tool.

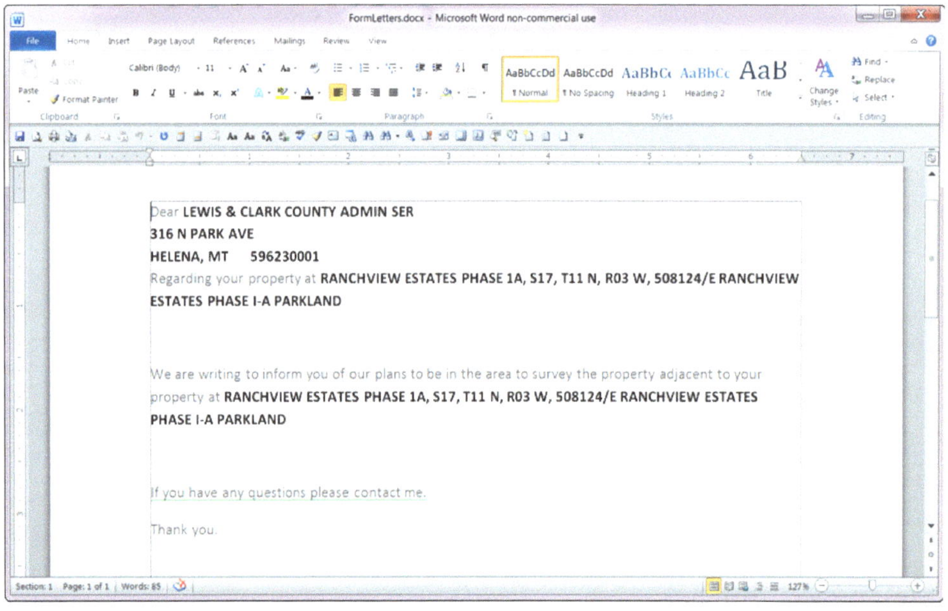

Dear **LEWIS & CLARK COUNTY ADMIN SER**
316 N PARK AVE
HELENA, MT 596230001
Regarding your property at **RANCHVIEW ESTATES PHASE 1A, S17, T11 N, R03 W, 508124/E RANCHVIEW ESTATES PHASE I-A PARKLAND**

We are writing to inform you of our plans to be in the area to survey the property adjacent to your property at **RANCHVIEW ESTATES PHASE 1A, S17, T11 N, R03 W, 508124/E RANCHVIEW ESTATES PHASE I-A PARKLAND**

If you have any questions please contact me.

Thank you.

FIGURE 89 FORM LETTER WITH DATA DISPLAYED IN PREVIEW

PART 5 GIS IN THE FIELD

In this part we discuss some ways that GIS helps the surveyor in the field and how field surveying is used to create and maintain GIS data.

CHAPTER 27 MOBILE ACCESS TO LAND SURVEY INFORMATION

Surveyors often need information from existing records while in the field. Existing records for things such as surveys, control points, land ownership, floodplain, and others are needed while in the field in order for a surveyor to help plan and execute a survey project. What things are available, where things are, how they fit together with other things on the ground, and other relevant bits of information regarding these things, such as who owns the property, what a property corner is made of, the elevation of a nearby benchmark, etc. are examples of the types of existing information that a surveyor may need while in the field. For example, to start a boundary survey for a property, the surveyor must find where the property is located, go to it, know what the dimensions are supposed to be, finding the property corners if any exist, and find nearby corners. Therefore, all surveys begin with a bit of research into the existing records, making copies of those records, and interpreting those records to understand where the survey is located and oriented in the world as well as how things fit together.

Whether existing records are in paper or other hardcopy form or digital, taking records into the field typically requires making a copy (either paper or digital) prior to leaving the office. Sometimes records are missed or forgotten or other information may become apparent while in the field which requires additional research. Wouldn't it be convenient to be able to do instantaneous research while in the field? With smartphone technology, teamed with web services and GIS, mobile access to information is now feasible and fairly simple to set up and implement. Smartphones can access the World Wide Web wherever phone service is available. Additionally, the location of the phone can be used to tailor information services appropriate for that location.

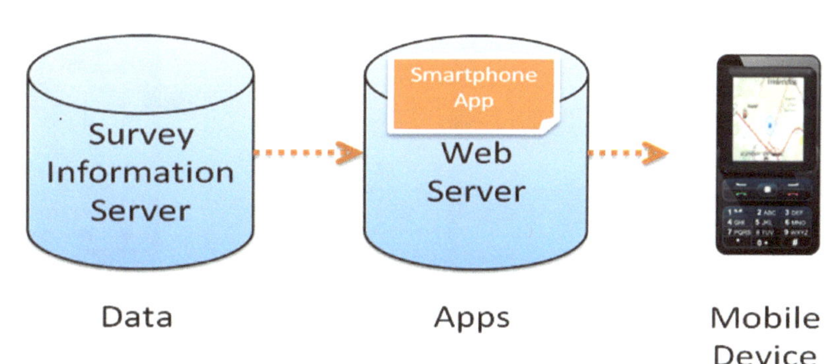

FIGURE 90 MOBILE LAND INFORMATION SYSTEM ARCHITECTURE

A Land Survey Information System is an information system of data of interest to surveyors. Since most of what surveyors do is location based, it makes sense that surveyors need location–based services as part of their work. A *mobile* Land Survey Information System (mLSIS) is one that serves the data such that it can be consumed by mobile devices. The ingredients necessary to make this a reality are digital data, web services, applications that run on a smart phone and smartphone type of mobile devices (figure 90).

An mLSIS can provide the surveyor with access to existing spatial data without having to download or copy large data sets. With mLSIS a surveyor can have location-aware access to relevant documents, and to data that are displayed in geographic space in context with other spatial data. Imagine driving down the road with a mobile phone that tells you where you are, who owns the property to the left and to the right of you, how far away the nearest geodetic control point is, shows an aerial photograph of the area, and calculates the driving route to the property destination. All this is within easy reach today.

Geographic Information Systems facilitate storing, querying, and accessing all sorts of spatial information including those data that surveyors require to do their work. As governments build geographic databases and make those data available to the public through web services, some of the information that surveyors require is there for the taking, although only a little survey information such as control points and survey records are there yet.

Let's take a look at a couple of examples of mobile surveyor information systems applications that are in use today.

MOBILE ACCESS TO LAND OWNERSHIP INFORMATION.

(from http://www.gcs-research.com, 2012)

GCS Research [of Missoula Montana] *developed an application using an Android-enabled smartphone to create a location-based application to access parcel information. The Parcel App (figure 91) allows smartphone users to access real estate information remotely. Using the built-in GPS, users can pinpoint their current location in the state and download parcel information.* The app works in Montana and

FIGURE 91 PARCEL APP BY GCS RESEARCH

Vermont where there are state-wide cadastral databases, and San Diego and Denver where there are city-wide cadastral databases published as map services.

The Parcel Apps allow smartphone users to access real estate information remotely from anywhere in the world or while standing on a property in the respective jurisdiction. Users can rely on the phone's built-in GPS to pinpoint their current location in the state, or they can zoom into a mapping interface to select an area of interest by "double-tapping" the touch screen. Within seconds, the application retrieves location-specific property information for the selected site from two web-enabled databases.

The first is a parcel boundary file maintained in a cadastral database. This cadastral information is managed in an ArcSDE geodatabase and served over the web by ESRI's ArcGIS Server solution. The smartphone application traces the property boundaries in red on the phone's map display. Next, the application accesses the cadastral system where it retrieves property details such as owner name, parcel legal description, acreage, assessed value, and zoning code. The mobile mash-up leverages the Android mobile operation systems to deliver an easy-to-use, personal solution for anyone wanting to have the information at their fingertips.

Accuracy of the parcel mapping combined with the accuracy of the GPS or mobile device positioning contribute to the uncertainty of the location of the property boundary. If other data are overlaid or base maps are used, then the inaccuracies of those layers also contribute to the uncertainty of the location of the boundaries. These uncertainties can and do vary from a few feet (less than a meter) to more than one hundred feet (30 meters). Thus, while these methods are helpful to ascertain who owns a property the exact location of the extents of the ownership are typically not reliable enough for boundary determinations. Although changes in GPS and other real-time positioning technologies and improvements in spatial accuracy of GIS mapping have improved mapping accuracies, the expectation and the ability to achieve even higher accuracy could change in the future. However, with today's technology this is not impossible to achieve, it is just not yet economically sensible to strive for because the techniques are costly and time consuming while the return on investment is low, primarily due to low demand for this level of accuracy.

MOBILE ACCESS TO EXISTING SURVEY CONTROL POINT INFORMATION.

Service New Brunswick Control Point Finder Application

Service New Brunswick (SNB) is the provincial government services department for the province of New Brunswick Canada. SNB's GIS division maintains a database of High Precision Control Points (HPN) which it makes available to the public through an online map service. The map service is accessible via the SNB website.

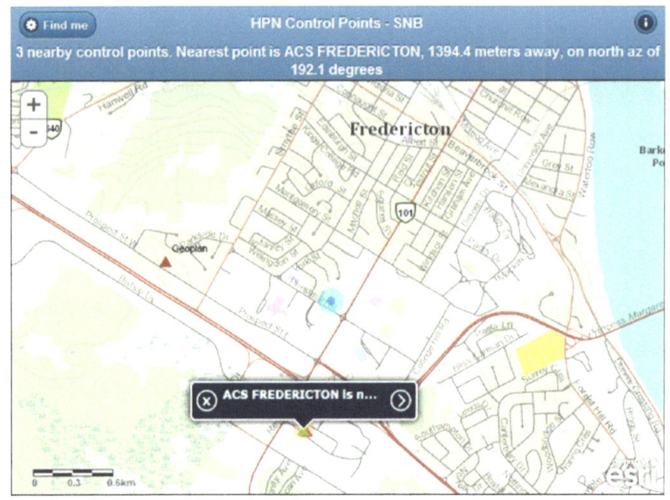

FIGURE 92 SERVICE NEW BRUNSWICK MOBILE CONTROL POINT FINDER APPLICATION

To demonstrate how survey information could be accessed while in the field, I created a mobile application using the SNB HPN map service. This mobile application will find the user's location; create a background map for location context, then display the SNB HPN points on the map. The application will also find the control point nearest the user and update that information as the user moves. Additional information about a particular point is presented in a popup when the user clicks on a point.

I built this app using data and services that are readily available: the Service New Brunswick High Precision Network control point map service, ESRI base maps map service, ESRI JavaScript API, the HTML5 geolocation service, and JavaScript. The control point finder app (figure 92) is a combination of HTML code and JavaScript which connects to the services, bundles them into presentation form and provides some basic functionality such as compiling a list of control points within the map extents, formatting which information will appear in a popup (figure 93), programming the application to move the map as the user moves, and calculating

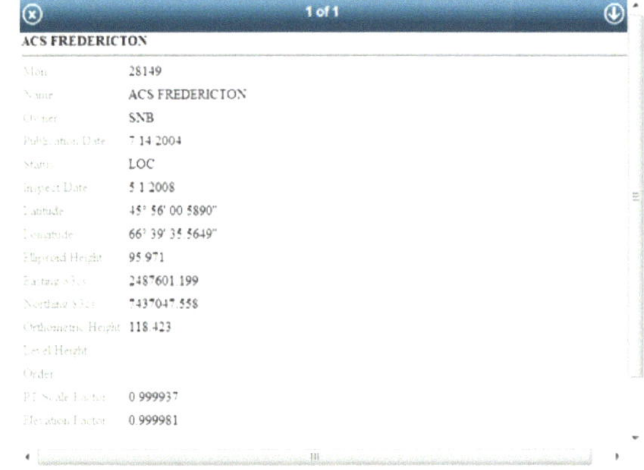

FIGURE 93 CONTROL POINT POPUP INFORMATION

the distance and direction to the nearest control point.

This code can be modified to support other control point map services such as the Multi-state Control Point Database of Idaho and Montana (http://mcpd.mt.gov/MCPD), as shown in figure 94. Others have created similar apps for accessing the National Geodetic Survey control points.

These examples demonstrate how surveyors can access existing information while in the field without having to copy data or datasets. The mLSIS dynamically selects subsets of the data that are appropriate to the location and data layers of interest to the surveyor.

The advantages of a mobile Land Surveyor Information System are:

- No need to go to the courthouse for research.
- No need to copy data.
- Research is done *real-time.*
- Research is done while in the field.
- Updates are automatic.
- Application *knows* your location and automatically pulls information pertinent to your location.
- Can help you navigate to features, making things easier to find.

With a bit more programming surveyors can also *contribute* information to online databases *while in the field*. For example, a surveyor could connect to an online database of corner records to search for corners *or* to create a new corner record while in the field, then post the new record to the online database.

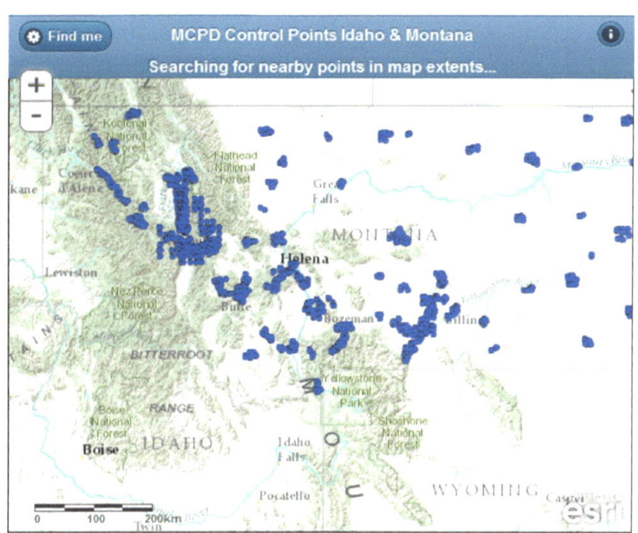

FIGURE 94 MONTANA-IDAHO CONTROL POINT FINDER MOBILE APP

The limiting factor is the availability of data that surveyors need. As long as the data are digitized they can be served to the public via the World Wide Web. If those data are spatially enabled by creating a graphic that is geo-referenced, then a *map* service can present the data in a spatial context. Thus, for example, an online database of corner records served through a map service could be used in to help surveyors find corners and their record information while in the field.

Mobile Land Survey Information Services can empower surveyors with quick access to information they need while in the field, while also providing spatial context to understand those data. This real-time information flow can go both ways allowing surveyors to be more efficient and more effective.

CORNER SEARCHES

A surveyor's search for property corners can be facilitated when property line information is loaded into a GPS unit as shown in figure 95. Property boundary GIS may be available from local governments or one may create the boundaries in the office from record data when necessary. The property boundary corners can then be used to identify a point coordinate for the probable location of the property corner monument. Rarely is the case that the GIS coordinate is the precise location of the property corner, but the GIS coordinate can narrow the search radius and help to speed up finding the actual corner in the field.

Surveyors in the western United States can obtain the approximate corner coordinates from the GCDB to input into a GPS as an aid to searching for corners in the field. The GCDB provides coordinates and error estimates of those coordinates (easting error and northing error).

Depending on the capability of your GPS, you may need to hand enter the coordinates to your GPS or you may be able to upload a file of the GCDB coordinates to your GPS.

1. Hand enter the GCDB coordinates into the GPS, then use the GPS unit's Go To or Navigate (to) options to generate a distance and direction to the corner from the current GPS location.

2. Download the GCDB coordinates to a file, convert the coordinates file to a format that your GPS can use, load the GCDB data into your GPS. Depending on the type of GPS equipment you have, you may see the GCDB points on the display (possibly with other background map data), and you may be able to see the GCDB information such as the point ID and coordinate reliability. Most importantly, you should now be able to navigate to the PLSS points based on the coordinates that are now loaded into your GPS.

GCDB points shown on GPS

GCDB point IDs info on GPS

GPS navigation to GCDB point

FIGURE 95 HANDHELD GPS WITH GCDB POINTS LOADED FOR FINDING PLSS CORNERS.

CHAPTER 28 FIELD MAPPING FOR GIS

Field mapping for GIS may be done for a variety of reasons such as inventory purposes, preliminary design, resource management, and recreation resource mapping.

The purpose of field mapping for GIS is to collect data that can only be collected or updated in the field. GPS field mapping provides the means to obtain a location and to identify what a feature is. Additional information may also be collected about individual features which may be used in GIS as attributes. For example one may wish to map the location of new water infrastructure as it is built. The features to map may include manholes, water lines, junctions, and valves. The water lines would be mapped as line features and the other features as point objects. Each feature may have attributes associated with them that one collects in the field such as the pipe material and diameter, direction of flow valve type, installation date, etc. The attributes become tabular data in the GIS database and may be used for cartographic display or for querying the database or used in analysis.

The questions to ask are what should be mapped (features), what characteristics of the features should be collected in the field (note that if tabular data already exists for the features, then it may not be necessary to collect that in the field), and how will the data be used (because this dictates the best way to map the things)?

Prior to entering the field a few things must be done. The first thing to do is to determine the spatial accuracy needed for the features that you plan to map because this will determine what GPS equipment and field procedures to use. GIS mapping is typically not done to very high spatial accuracy standards compared to most land surveying accuracies. Usually, a few decimeters to a few meters is adequate depending on the intended use of the data. The second thing to do is to create a data dictionary (mapping template) that simplifies field mapping and imposes uniformity of the data structure and format. The data dictionary is designed and loaded into the GPS, tested then modified as needed. The final thing to do before going into the field is GPS mission planning which is to see the satellite availability for the time and location where and when one plans to map. It is important to know if sufficient satellites will be available to cover the project area during the time you plan to map, in order to achieve the desired accuracy.

For large projects GIS is helpful for planning, managing, and communicating the resources and phases of the project as described in the chapter on GIS for Managing a Project. The Trail Mapping Project listed below is an example of this.

MAPPING TRAILS

Trails are mapped for resource management, inventory, planning, and communication purposes. An agency or recreation group may wish to understand the location and length of their trails, and they may wish to publish maps in paper or web formats, and they may want an inventory of features along their trails such as points of interest, the number of bridges, kiosks, interpretive signs, directional signs, and other supporting objects which must be maintained over time.

For the most part, a precise location survey is not necessary as the spatial accuracy needs, which are dictated by the intended use of the mapping, may vary from a few meters to 100 meters.

Hiking, biking, horse riding, skiing, and other recreational trails can be mapped using GPS. For example, I worked a project for a national forest to map all their recreation trails – around 900 miles (more than 1400 km) of trails.

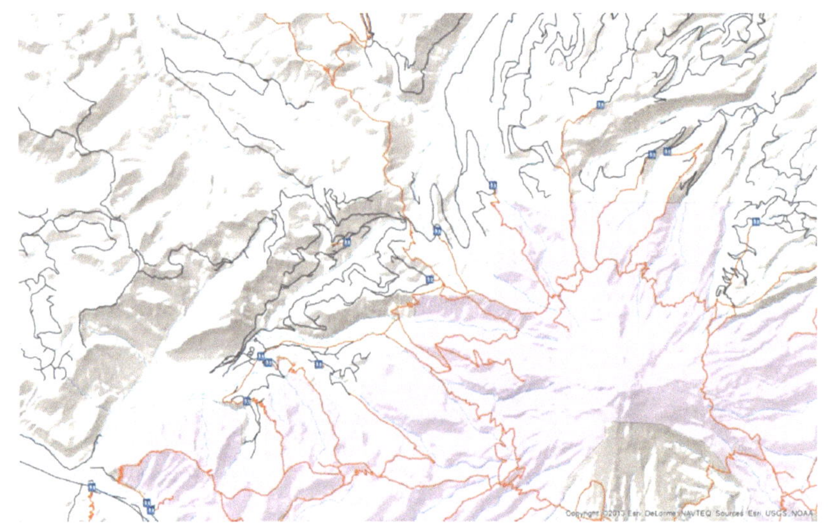

FIGURE 96 PROJECT PARAMETERS – TRAILS TO MAP, TRAIL HEADS, ROAD ACCESS

The features to map were trails, trailheads, trail junctions with roads or other trails, stream crossings, bridges, and any points of interest such as water falls, or exceptional views. The spatial accuracy required was 3-5 meters, and the contract stipulated 1-second GPS epochs for all features. This GPS specification is a time based interval. For trails mapped as linear features it usually makes more sense to use a distance interval such as 1-2 meter distance between recorded GPS positions. This prevents "stacking" GPS positions in a small area such as when one stops walking. Stacked GPS fixes generates unnecessary positions for very short line segments and since the GPS coordinates can vary even when the GPS unit is held stationary, those line segments will become a twisted knot which is a messy line while it increases the file size without providing any additional useful information. Time intervals are good for point features because the additional positions fixes are averaged to a single position. The attributes collected for these features were trail name and trail number for the trails and trail heads.

GIS helped to layout the project parameters as shown in figure 96. GIS was used to plan the field work. The trails were scattered throughout the forest, and some trails had road access while other trails could only be accessed from other trails. Therefore careful planned was required in order to map the trail efficiently. Daily plans of who would map which trails, where they would be dropped off in the morning and where they would be picked up at the end of the day (often a different location). Detailed information regarding each trail was available in a spreadsheet which was joined to a GIS feature set of approximate trail locations (mapped from a variety of sources which were mostly digitized from USGS topographical maps). Project management data were added to the spreadsheet to track mapping assignments, dates when the GPS mapping was completed, and when the GPS data were processed to differentially correct it.

GIS allowed quick estimates of the number of miles one could map without or without back tracking, and also provided driving directions to the trail head. GIS helped with the daily planning by providing a graphical perspective to assess trail connectivity, access points for drop-off and pick-up, estimate travel times, get driving directions to trail heads, see which trail segments to remap under better GPS conditions, and understand which direction to hike for maximizing downhill hiking for speed and efficiency.

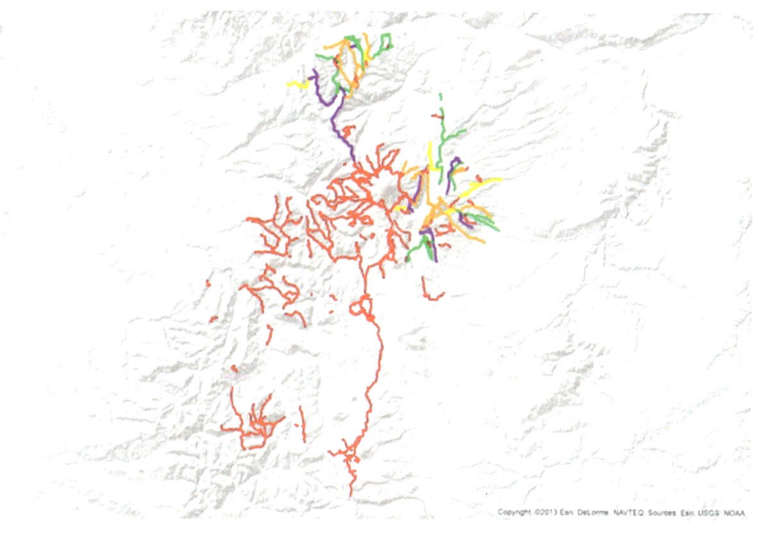

FIGURE 97 TRAILS MAPPED BY PERSON (RED TRAILS WERE NOT YET MAPPED)

The daily crew assignments were reviewed with the entire crew and PDFs of daily assignment logs and maps were emailed to the client and the home office. Each crew member got paper copies of everyone's assignment, along with maps for how to get to the trails, and maps of the trail routes to map. All the trails and roads data were also loaded into the GPS as reference data to aid in orientation.

As trails were completed maps were created to show where work was done as shown in figure 97. Overall progress maps showed which trails were done and by whom. Tracking who mapped which trails helped when there were any questions arose such as how much of a trail was completed or if the trail no longer existed on the ground. Progress reports for number of miles mapped could quickly summed using GIS by opening the table for the mapped trails features and summing their lengths.

PART 6 SPATIAL ACCURACY

Spatial accuracy is a paramount concern for land surveyors. In this part we discuss a variety of aspects of spatial accuracy of GIS data, beginning with a discussion of accuracy and precision in GIS mapping, then we discuss how to test the spatial accuracy of GIS data and how to improve the spatial accuracy of data when creating the data as well how to improve the spatial accuracy of existing GIS data such as parcel lines.

CHAPTER 29 GIS ACCURACY

ACCURACY IS RELATIVE

Accuracy of GIS data are relative to the type of data to be digitized, the intended use for the data, the time and money available to convert data and the accuracy of the source information. Accuracy requirements may vary by location. For instance an agency may require greater positional accuracy for parcels in an urban area than for rural areas with large tracts. Since GIS data accuracy requirements are relative not absolute, the absolute accuracy itself is not as important as reporting what the accuracy of the data are. Data spatial accuracy reporting is typically in the metadata report. The metadata report describes other important data conversion information such as source documents, intended purpose for the data, what conversion methods (such as GPS, on-screen digitizing and COGO) were used, accuracy standards applied if any. This accuracy report is similar to what surveyors typically put into the narrative portion of a plat or survey.

(for a complete description of metadata standards, visit www.fgdc.gov/metadata)

FIGURE 98 ZONING BASED ON PLSS, PARCELS AND HYDROGRAPHY

While spatial accuracy is often a high concern for surveyors, there are many instances in GIS where the spatial accuracy is not paramount because the intended uses of the GIS data don't require high spatial accuracy. For example, if one is only interested in having a graphic index to parcel information then the spatial accuracy of the parcel lines is not a large concern; one may simply click anywhere on the parcel to access the tabular data. However, because GIS can also show the juxtaposition of the parcel layer with other geographic data, the accuracy of the various datasets, including the parcels, does become important. Figure 98 shows the parcel layer with zoning information. In this example, the zoning boundaries (as is typical) follow public lands survey system (PLSS) section lines, parcel lines, rights-of-way and hydrography boundaries. Zoning therefore depends on the existence of other layers for its accuracy. Here it is more important that a zoning boundary be spatially consistent with a parcel than within a certain absolute distance (± x feet). That is, accuracy relative to another GIS layer is more important than absolute accuracy.

SURVEY EXAMPLE

Now look at an example relevant to a surveyor. Your client asks you for an estimate to perform a survey of his or her lot. Because you have access to the county GIS online, you can begin your research right there at your desk while still talking to your client. You pull up the parcel data and the county survey index (see figure 99). The parcels are gray lines, and the survey index numbers are blue text with CS prefix). You can zoom and pan to a particular parcel layer, then click on the parcel to obtain tax assessor record of the parcels. The GIS gives you the parcel information, including parcel size and ownership. You have immediately confirmed that, indeed, your client owns this property and you readily see where the property is located. You then select the survey information layer and again click on the parcel on the screen. Now the GIS returns a list of the surveys that have been done on or around on that property.

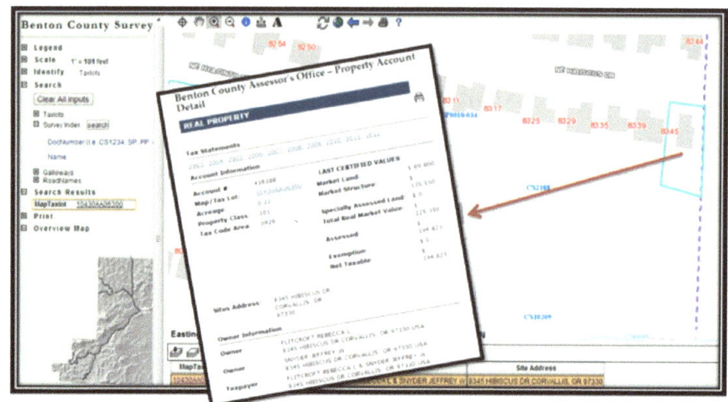

FIGURE 99 CLICKING ON A PARCEL TO ACCESS TABULAR DATA

You are feeling good that there are at least four surveys that included this parcel, but you want to examine the most recent one, in this case CS5669. Figure 100 shows the approximate extents of CS5669 (highlighted in light blue), so you hit the hot link button and up pops the scanned image of CS5669. You see that this survey was performed in 1974 by a surveyor of good repute and that all the pins were set or found. You do a few

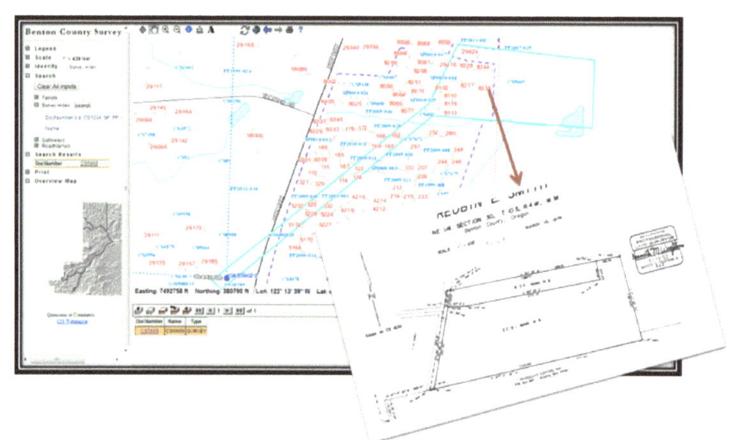

FIGURE 100 ONLINE COUNTY SURVEY INDEX WITH LINKS TO SCANNED SURVEY

more checks of the surrounding parcels and surveys and check the plats, and in a few minutes you are ready to give an estimate.

How accurate were these parcel lines and surveys mapped in the GIS? You do not know off-hand. You could check the metadata, but does it really matter? You were able to do the majority of your research right there at your desk and provide your client with an estimate very quickly because you had a quick spatial index and access to the public records. You did not have to drive to the county courthouse to look up and pull the surveys and make copies. You had all the surveys at your fingertips and could even see how the surveys sit with respect to your client's property as well as with surveys of other properties. GIS served you well for this intended purpose, irrespective of its spatial accuracy. This is the power of GIS—to access information quickly and be able to see the spatial juxtaposition of various themes (such as parcels, zoning, and surveys).

ACCURACY OF GIS DATA COLLECTED BY GPS

A common mistake among GIS folks when creating GIS data using GPS is to rely completely on the GPS user's manual to tell them what the accuracy of their data will be if they follow certain procedures. For instance, the user's manual may state that in order to achieve sub-meter accuracy you must have a PDOP of 4, a SNR of 4, track at least 5 satellites and perform a differential correction. Many GIS folks read the reverse in that statement and assume that if the conditions are met then the data will be sub-meter. That is similar to thinking that if you used a steel tape that's graduated to 1/100's of a foot means that all your measurements with that tape will be accurate to 1/100th foot regardless of how you do the measuring. But, of course this is not necessarily true. There are many factors that affect the quality and accuracy of field measurements. Some of those factors are knowable and quantifiable, some can be controlled, and some factors are beyond human control or knowledge. Another issue, with respect to GPS coordinates is the difference between relative accuracy and absolute accuracy. Most resource grade GPS software will provide an error report for a dataset. Yet, typically, what they are reporting is the relative error of the measurements not the absolute error. That is a report stating an accuracy of +/- 1 meter is reporting that the error of that set of measurements is within a meter of each other. That does not necessarily mean that those coordinates are within a meter of the true

> - FGDC accuracy standards: http://www.fgdc.gov/standards/status/sub1_3.html
>
> "The NSSDA is intended to replace the 1947 National Map Accuracy Standard (NMAS). The applicability of NMAS is limited to graphic maps, as accuracy is defined by map scale. The NSSDA was developed to report accuracy of digital geospatial data that is not constrained by scale."

(absolute) coordinate. In order to test the absolute error, points with known higher accuracy coordinates must be observed in order to estimate the probable error.

GIS data may be created from GPS, by digitizing source documents, performing field surveys, aerial mapping, address matching, and other methods. Regardless of the way, a GIS layer is obtained, or acquired; there will be positional errors in the dataset. Determining the magnitude of the positional (or location) error is important because the usability of the dataset may be dependent upon its spatial accuracy. Metadata, often referred to as the data about the data, is essential for providing potential users with the information needed to determine a GIS dataset's usability for an intended purpose. One of the Metadata content components is a statement of spatial accuracy. Spatial accuracy is probably the most important issue that surveyors focus on most when criticizing GIS. However, although spatial accuracy is a big concern to surveyors, it's important to keep in mind that it is not necessarily the most important issue to all GIS users. For instance, when emergency responders need information on house locations to aid in developing evacuation plans, they're priority is on getting the information as *quickly* as possible. They do not care whether the houses were mapped to an accuracy of +/-10 meters, or 30 meters or whatever. Additionally, wildlife biologists using GIS to study land use patterns of a watershed may use small scale mapping as low as 1:100,000 which, according to National Mapping Accuracy Standards would require accuracy of +/ 50 meters – which is large error to a surveyor, but inconsequential for a biologist.

While the importance of the magnitude of the spatial accuracy may vary, reporting what that value is very important is important because it informs the potential data user what that spatial accuracy is. With that information the user can determine whether or not a dataset will work for his or her intended use. Although guidelines for spatial and/or mapping accuracy do exist (see sidebar), the data creator may or may not choose to follow those guidelines. In any case, when data are mapped or converted there is usually some kind of mapping or spatial accuracy goal that the project must achieve. Here are some methods for testing and validating the accuracy of GIS datasets.

TYPES OF GIS DATA

GIS data can be described as points, lines, areas, and raster data. Each of these data types has its unique requirements for spatial accuracy testing. However, for all types of data there are some common considerations to observe. In order to determine the spatial accuracy of a GIS dataset the following must be considered: determine what to test, decide how to test it, (procedure, sample size, sample method), analyze the sample data, and report the results.

DATASET	TEST FEATURES
AERIAL PHOTOGRAPHY	ROAD INTERSECTIONS, FENCE CORNERS, MANHOLES
PARCEL LINES	PROPERTY CORNERS, RIGHT-OF-WAY,
CONTOURS	VERTICAL AND HORIZONTAL LOCATION OF SIDEWALK CORNERS, ROCK OUTCROPS, EXISTING BENCHMARKS

29.1 HOW TO DETERMINE THE SPATIAL ACCURACY OF GIS DATASET

The procedures to determine the spatial accuracy of a dataset are basically, to identify distinct point features in the test dataset; obtain the coordinates of those points from the dataset; perform field measurements or measurements in a comparison dataset of the same features (independent source); compare the two sets of measurements; and report the results.

There are two methods for determining the spatial accuracy of a dataset, perform new field measurements or perform comparative measurements against another (independent) dataset that is known to have a higher spatial accuracy than the test dataset. To determine the spatial accuracy of a dataset, one must select distinct points on the dataset that can be located on the ground or located in a reference dataset. Some examples of test features for spatial accuracy testing are shown in the chart below.

The number of sample points to measure should consist of a sufficient number of points to provide a statistically reliable level of confidence in the determination. The Federal Geographic Data Committee (FGDC) recommendation (Geospatial Positioning Accuracy Standards Part 3: National Standard for Spatial Data Accuracy) is to measure a minimum of twenty (20) test points. If twenty test points are used, then one measurement can fail the test at the 95% percent confidence level for a given threshold. Other numbers of test points (more or less), may be used depending on the number of features in the dataset, scale of the data, availability of the test points, field access issues, etc.

The geographic distribution of the test points should correspond with the distribution of the features within the dataset, unless other factors indicate otherwise. Other factors may be such things as physical and legal accessibility of the selected points for field measurements, which features or areas in the dataset are more important, the location and distribution of features in the dataset.

REPORTING

Reporting conventions vary according to the standard that one is reporting to. Some examples are listed below. Reporting spatial accuracy consists of a statement of the numerical accuracy of the

SCALE 1INCH = X	FEET	HORIZONTAL ACCURACY (FT)	VERTICAL ACCURACY (FT)
1:1200	100	3.33	[CONTOUR INTERVAL]*0.5
1:2400	200	6.67	[CONTOUR INTERVAL]*0.5
1:4800	400	13.33	[CONTOUR INTERVAL]*0.5
1;12000	1000	33.33	[CONTOUR INTERVAL]*0.5
1:24000	2000	40.0	[CONTOUR INTERVAL]*0.5

dataset at the 95% confidence level, and/or the standard met (or failed). In addition to a statement of the standard and/or numerical value, one should also include a description of how the test measurements were made, and the methods used to determine the spatial accuracy. Numerical accuracy should be reported in ground units. Depending on requirements, there may be statements of horizontal accuracy and/or vertical accuracy. Also, note that spatial accuracy certification may be part of the data compilation, and not necessarily a post-compilation accuracy test. Statements regarding compilation are usually similar to statements for testing but for the substitution of the word compiled for the word tested.

A simple numerical statement of spatial accuracy may read such as the FGDC statement
Tested ____ (meters, feet) horizontal accuracy at 95% confidence level
(Geospatial Positioning Accuracy Standard, Part 1, Reporting Methodology,
FGDC-STD-007.1-1998)
Quantitative statements such as National Map Accuracy Standards or ASPRS are listed below.
National Map Accuracy Standards
This map complies with National Map Accuracy Standards of 1947 for horizontal [or vertical or horizontal and vertical] accuracy.

National Map Accuracy standards for large scale mapping

ASPRS:
These data were checked for accuracy and found to conform to the ASPRS standard for class (1., 2., 3.) Accuracy

ASPRS Accuracy Standards for Large-Scale Maps Class 1 horizontal (x or y) limiting RMSE for various map scales at ground scale for metric units:

MAP SCALE	CLASS 1 PLANIMETRIC ACCURACY LIMITING RMSE (METERS)
1:50	0.0125
1:100	0.025
1:200	0.050
1:500	0.125
1:1,000	0.25
1:2,000	0.50
1:4,000	1.00
1:5,000	1.25
1:10,000	2.50
1:20,000	5.00

METADATA STATEMENTS:

Although there are a variety of forms for metadata the most commonly accepted form is the FGDC format. The FGDC for guidelines on how to report the spatial accuracy of a digital dataset in a (FGDC compliant) metadata report (section 2) are shown below.

(Data_Quality_Information/Positional_Accuracy/Horizontal_Positional_Accuracy /Horizontal_Positional_Accuracy_Assessment/Horizontal_Positional_Accuracy_V alue)
and/or
(Data_Quality_Information/Positional_Accuracy/Vertical_Positional_Accuracy/V ertical_Positional_Accuracy_Assessment/Vertical_Positional_Accuracy_Value)
Enter the text "National Standard for Spatial Data Accuracy" for these metadata elements (Federal
Geographic Data Committee, 1998, Section 2), as appropriate to dataset spatial characteristics:
(Data_Quality_Information/Positional_Accuracy/Horizontal_Positional_Accuracy /Horizontal_Positional_Accuracy_Assessment/Horizontal_Positional_Accuracy_E xplanation)
and/or
(Data_Quality_Information/Positional_Accuracy/Vertical_Positional_Accuracy/V ertical_Positional_Accuracy_Assessment/Vertical_Positional_Accuracy_Explanati on)

As the GIS community and the public come to appreciate the importance of reliable spatial accuracy statements, surveyors will increasingly be called upon to certify the accuracy of GIS data.

CHAPTER 30 TESTING SPATIAL ACCURACY

There is a variety of ways to perform the spatial accuracy testing. Typically the procedures are to use some sort of measurement or test that is extrinsic to the dataset, i.e. an independent data source or computation. Independent sources should be of a higher accuracy than the dataset to be tested. Some examples are existing digital or hard copy map data, GPS, or terrestrial survey data. In lieu of extrinsic data, estimates can be computed from intrinsic sources such as knowledge of the accuracy of the source document, map registration and digitizing accuracy (based on scale and methods used), etc. But, creating independent measurements assure the highest reliability of the accuracy determination.

The methods selected should depend on the objectives and the availability of existing data. If the GIS objective is to fit a dataset into other existing (higher accuracy) data, then the new set may be tested against the existing data. For example, the State of Texas created a state-wide GIS layer of public roads with the requirement that it correlate with existing digital orthorectified photography (see Figure 101 GPS road centerline vs. aerial photography for an example GIS road centerline created from GPS mapping and overlaid on an orthophotograph). In that case, the accuracy of GPS data can be tested by overlaying the road network on the photography, then measuring, on-screen, the difference between the image of a road segment and the GPS road segment. For instance, a road intersection on the photography and the GPS centerline would be the test. The distance between them would be a single sample. Natural resources data, such as vegetation coverages, are more difficult to test because such datasets do not describe well defined points.

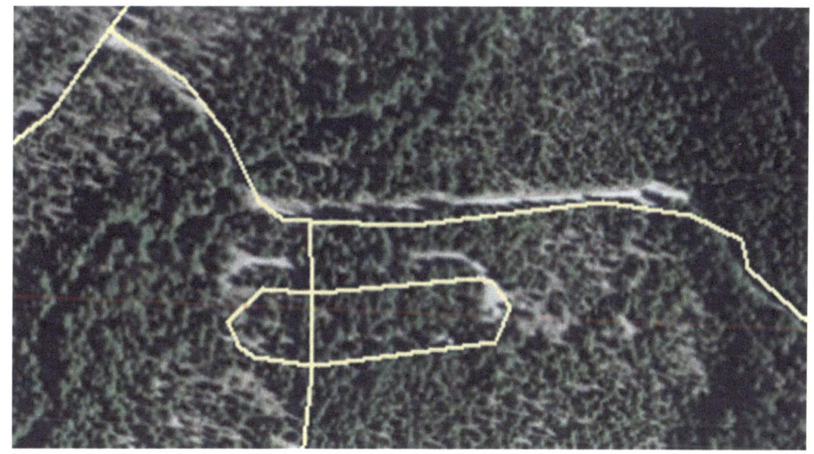

FIGURE 101 GPS ROAD CENTERLINE VS. AERIAL PHOTOGRAPHY

Another way to perform testing is to identify points in the dataset that can be physically measured in the field, then measuring those samples with higher accuracy methods. For example, if the accuracy of a manhole inventory was being checked, then a sample set of manholes would be re-measured with higher accuracy equipment and methods. If the GIS requirement was to map those manholes to a one-meter level of accuracy, then a sample set of manholes should be tested using methods that yield better than a meter accuracy. The difference between the original measurements and the test measurements will be the accuracy of the manhole dataset.

SOME ACCURACY TESTING SAMPLE METHODS:

Points are the simplest GIS features to test. Points have only a location (i.e. coordinate), so the method would be to obtain the coordinates of the test point, then compare that to the coordinates of the same point as defined by a higher accuracy source (such as GPS). Lines and areas, however, are more complex features to test, because their geometry is more complex. The consideration for the more complex geometry datasets is to test the geometry as well as the coordinates of discrete points. Testing lines, for example, requires testing the accuracy of the end points of the lines such as at road intersections, and the geometry of the line between the endpoints. The geometry is how the line behaves between the end points - does it go in a straight line, does it curve left or right, etc. Testing the end points of the lines is the same as testing a simple point. However, testing the accuracy of the geometry of the line requires obtaining a coordinate for some point or points along that geometry. Since a road centerline created from parametric modeling (i.e. typing in the curve data) would create a more mathematically correct geometry, than a GPS representation of that same curve, the difference can be difficult to differentiate. A sharp curve would require more intermediate GPS points than a flatter curve.

Testing the more complex geometry of a polygon requires finding discrete points on the edges of the polygon, such as vertices, which can be correlated to similar points on the dataset of higher accuracy. If the polygon were a parcel boundary, for instance, one of the checks for spatial accuracy would be to test a corner of the parcel for accuracy.

ACCURACY TESTING SAMPLE SIZE

An important consideration when testing spatial accuracy is selecting an appropriate size sample set. There must be a large enough sample size to produce a statistically valid result. On the other hand the sample size is constrained by the cost and the time required to perform the sampling. Surveyors well understand the value of redundant measurements, but there is a point after which the value of more measurements diminishes. The balance between the number of measurements and the value of the measurements is unique for each project and should be dealt with individually. Some of the factors to consider are the size of the dataset (is it 10 points or 10,000 points), the geographic distribution of the data, and the importance of the spatial accuracy. If the spatial accuracy is of high importance then that may be incentive to test a large sample set than the minimum required. There are statistical programs available for calculating sampling sizes for known and unknown dataset sizes.

For photography, a common sample size is twenty points per image, distributed randomly throughout the image. Line data, such as river or road networks could be tested a few of different ways. A certain portion of the total length of the dataset could be tested, or a percentage of characteristics, such as junctions or angle points could be tested. Additionally a random set (such as 1%) of the points along the network could be tested, or measurements could be taken at predetermined intervals (X distance or Y% along the network). For example, if a linear network were 100 kilometers long, then the sample set may consist of points every ½ km, or 1 km, along the route for a standard interval. Alternatively sample may be taken every 5%, or 20% along the way.

REPORTING ACCURACY TESTING RESULTS

One method of reporting the spatial accuracy is the FGDC Metadata Content Standard for Spatial Data Accuracy. The standard requires a quantitative and qualitative statement of accuracy. Additionally, most survey adjustment software and GPS software provide reports in a variety of proprietary formats (some are customizable). The important things that are generally helpful to the end user are a quantitative statement of accuracy and some information about how that was determined.

An (abridged) example is shown below, taken from the North Texas GIS Consortium Metadata for Pavement (http://www.ntgisc.org/warehouse/metadata/roadedge.html).

Positional_Accuracy Horizontal_Positional_Accuracy
Horizontal_Positional_Accuracy_Report: Horizontal positional accuracy is 1.0 meter defined by the root mean square error (RMSE) method. This requires that two-thirds of all photo-identifiable arc features fall within the stated accuracy of 1.0 meter and that 90 percent of all arcs must fall within twice the distance specified (i.e. 2.0 meters). A final inspection and acceptance process was completed by several North Texas Consortium members.
Quantitative_Horizontal_Positional_Accuracy_Assessment:
Horizontal_Positional_Accuracy_Value: 3.2ft
Horizontal_Positional_Accuracy_Explanation: Resolution as reported
Vertical_Positional_Accuracy
Vertical_Positional_Accuracy_Report: Vertical positional accuracy of 1 meter RMSE.

As the push to share geographic information increases, the need to verify and document the spatial accuracy of datasets becomes ever more important. Assessing the spatial accuracy of GIS data are a niche that surveyors are well suited for, and to which surveyors should lend their expertise.

CHAPTER 31 CERTIFYING SPATIAL ACCURACY OF GIS DATA

One important role that surveyors have in Geographic Information Systems is to certify the spatial accuracy of GIS data. GIS metadata standards require a statement about spatial accuracy. Usually those statements are rough unsubstantiated estimates. Generally, such rough estimates are sufficient to give the user or potential user of the data, an adequate idea about the utility of the data for one purpose or another. However, such statements are often mere guesses and do not provide any level of assurance of the spatial accuracy. Nevertheless, there are instances when a higher level of assurance of the spatial accuracy of a dataset is important. Examples of GIS datasets that might be certified would include datasets that are used for regulatory purposes such as flood plain maps; or that serve as a base layer from which other GIS layers are built, such as aerial photography. Additionally, datasets that are used to align other GIS layers or that are used as a basis to determine the spatial accuracy of other datasets; or datasets that are to be used for engineering design, should all be certified, so ensure that the accuracy is sufficient for the intended use.

To certify accuracy is to provide a higher level of assurance than to merely state accuracy. To certify is to attest authoritatively: to attest as being true or as represented or as meeting a standard; to inform with certainty: to assure. Anyone may make a statement regarding the accuracy of a GIS dataset, but only a surveyor has the training, experience, and thus the credentials to provide the weight of an authoritative assurance to a statement of spatial accuracy. Surveyors are experts at measuring, detecting measurement blunders, isolating errors, designing error detection procedures, and error analysis. Surveyors know how to gather sufficient measurement data to estimate the spatial error of a dataset, and how to estimate the magnitude and types of error inherent in the measurement equipment and the atmospheric effects on measurements and equipment. The surveyor also understands how to design redundant measurements, as well as how to analyze the measurement data to determine the accuracy of the measurements, and thus estimate the accuracy of a dataset.

There are two types of spatial accuracy certification: qualitative and quantitative. Qualitative certification is a statement that the spatial accuracy meets a particular standard, such as National Map Accuracy Standards, or Accuracy Standards for Large-Scale Maps by the American Society for Photogrammetry and Remote Sensing (ASPRS)., etc. A quantitative certification is just a statement of the numerical value of the accuracy (or error) of a dataset, without reference to a standard. Either type of certification may be called for or both may be used together, depending on the needs of the data developer, the client, the intended use, or standards for an industry. However it is good practice to always include a quantitative statement when stating whether (or not) a map or dataset conforms to a qualitative standard. Qualitative accuracy reports require that one determine the magnitude of error, i.e. to quantify the accuracy and compare that error against a standard to determine whether or not it meets that standard, or within which class- standard the accuracy falls.

WHERE TO FIND STANDARDS FOR SPATIAL ACCURACY.

Map and digital data spatial accuracy standards exist and anyone who has a mapping project may elect to apply an existing standard their dataset. When existing standards are insufficient or inapplicable, one may devised a threshold for accuracy based on the requirements for a particular use of a dataset. The more common and often referred to standards are the US National Map Accuracy Standards, the American Society of Photogrammetry and Remote Sensing Standards for Large Scale Maps, the Federal Geographic Data Committee standard for geospatial positioning, and the US Army Corps of Engineers standards manuals.

- US Geological Survey (USGS) National Mapping Program Standards
- United States National Map Accuracy Standards - defines accuracy standards for published maps, including horizontal and vertical accuracy, accuracy testing method, accuracy labeling on published maps, labeling when a map is an enlargement of another map, and basic information for map construction as to latitude and longitude boundaries. http://geography.usgs.gov/standards/

ASPRS

The American Society for Photogrammetry and Remote Sensing of http://www.asprs.org/ has draft standards that are useful for aerial photography and photogrammetry. These are standards are referenced often, and provide excellent guidelines for photography and photogrammetry projects

FEDERAL GEOGRAPHIC DATA COMMITTEE

The Federal Geographic Data Committee developed four standards for positioning that are of interest to the surveyor. The standards are for reporting methodology; geodetic control networks; spatial data accuracy; and architecture, engineering construction and facilities management (A/E/C). The aim of the FGDC standards is to provide consistency in determining and reporting spatial accuracy and, for certain categories of data, such as A/E/C, to provide accuracy thresholds for data. The standards may be found on the FGDC web site at the links listed below. These are very important standards for the geospatial community.

Geospatial Positioning Accuracy Standard, Part 1, Reporting Methodology, FGDC-STD-007.1-1998

Geospatial Positioning Accuracy Standard, Part 2, Geodetic Control Networks, FGDC-STD-007.2-1998

Geospatial Positioning Accuracy Standard, Part 3, National Standard for Spatial Data Accuracy, FGDC-STD-007.3-1998
(The U.S. Geological Survey has submitted a proposal to revise this standard)

Geospatial Positioning Accuracy Standard, Part 4: Architecture, Engineering Construction and Facilities Management, FGDC-STD-007.4-2002

Federal agencies and the GIS community are encouraged to follow FGDC standards for geospatial positioning accuracy. A portion of the FGDC standard for Positioning Accuracy is shown below.

Federal Geographic Data Committee FGDC-STD-007.3-1998
Geospatial Positioning Accuracy Standards
Part 3: National Standard for Spatial Data Accuracy
Objective
The National Standard for Spatial Data Accuracy (NSSDA) implements a statistical and testing methodology for estimating the positional accuracy of points on maps and in digital geospatial data, with respect to geo-referenced ground positions of higher accuracy.
Scope
The NSSDA applies to fully geo-referenced maps and digital geospatial data, in raster, point, or vector format, derived from sources such as aerial photographs, satellite imagery, and ground surveys. It provides a common language for reporting accuracy to facilitate the identification of spatial data for geographic applications.
This standard is classified as a Data Usability Standard by the Federal Geographic Data Committee
Standards Reference Model. A Data Usability Standard describes how to express "the applicability or essence of a dataset or data element" and includes "data quality, assessment, accuracy, and reporting or documentation standards" (FGDC, 1996, p. 8)
This standard does not define threshold accuracy values. Agencies are encouraged to establish thresholds for their product specifications and applications and for contracting purposes. Ultimately, users identify acceptable accuracies for their applications. Data and map producers must determine

what accuracy exists or is achievable for their data and report it according to NSSDA.

It is important to note that, in most instances, the FGDC Geospatial Accuracy standard does not describe what accuracy a dataset ought to have, rather, the standard describes procedures for determining the accuracy and a language for communicating the accuracy.

US ARMY CORPS OF ENGINEERS

The U.S. Army Corps of Engineers is responsible for developing and maintaining the A/E/C geospatial positional accuracy data standards for the Facilities Working Group of the Federal Geographic Data Committee. Address questions concerning the standards to: Headquarters, U.S. Army Corps of Engineers, ATTN: CECW-EP, 20 Massachusetts Avenue NW, Washington, D.C. 20314-1000.

http://www.usace.army.mil/usace-docs/eng-manuals/em1110-1-1000/toc.htm

CHAPTER 32 EVALUATING SPATIAL ACCURACY OF PARCEL DATA

There are various ways to evaluate the spatial accuracy of GIS layers. Here we describe a study performed by a land surveying/GIS firm to quantify the spatial accuracy of parcel data within a one square mile test area. Additional work was performed to

- Categorize the types of problems and issues discovered,
- Determine the probable cause of large errors,
- Determine the most economical process to improve the spatial accuracy where improvements were possible.
- Make improvements, where possible, to the spatial accuracy of those features that were out of compliance with the target accuracy (1 – 5 feet).
- Report on the cost to perform the accuracy assessment and the cost to improve the accuracy of the data.

A consulting surveyor working with a GIS technician performed an initial accuracy assessment of the parcel data by comparing parcel line work to the built environment visible on existing high resolution digital orthographically-rectified photography that was mapped at 1:1,200 scale. The spatial accuracy of the photography exceeded the National Map Accuracy Standard of 3.33 feet in most areas of the photograph and averaged around 1.5 feet in most places. Therefore, the photography was a reliable and useable basis for testing the accuracy of the parcels and allowed the majority of the testing to be performed in the office. The project was performed in two phases – Phase I was a spatial accuracy assessment on the original parcel data using a proscribed work

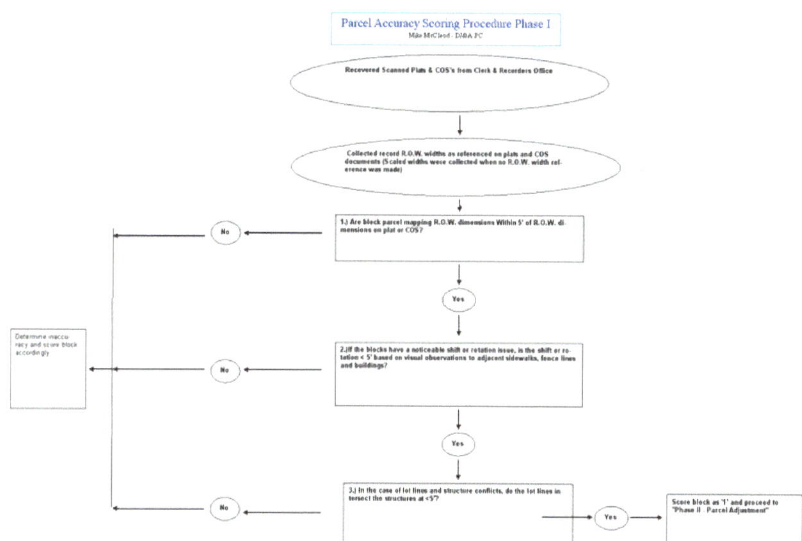

FIGURE 102 EXAMPLE FLOW CHART FOR PARCEL ACCURACY ASSESSMENT

flow outlined in figure 102. Phase II involved testing different methods to improve the spatial accuracy of those parcels.

PHASE I- ACCURACY ASSESSMENT

Phase I of the parcel adjustment process required development of a subjective numeric and color coded accuracy rating system to rank the spatial accuracy for each city block. The numeric accuracy rating was determined by overlaying and visually comparing the parcel data to the orthorectified aerial photographs. Visual observations for the planimetric features (fence lines, sidewalks, structures, edge of asphalt or curb lines of streets and roads) were the primary items used to reference parcel lot lines in determining a block rating classification. Random measurements between the parcel mapping lines and the record documents were also conducted and compared to record distances on plat maps and certificates of survey (COS). The parcel accuracy scoring procedures (Figure 102 Example flow chart for parcel accuracy assessment) details the steps taken to identify each of the blocks rated. If a block met all criteria, it was given a rating of 1. If the criteria were not met on any level, then a numerical rating of 2, 3, or 4 was assigned depending on the amount of linear discrepancy between existing parcel lines and the orthophotography. Areas where there was a lack of information necessary to quantify the error, were flagged with a 6 – for undetermined. When all blocks were completed, a map was developed to display the assigned rating for each block (Figure 103 Reliability of original parcels).

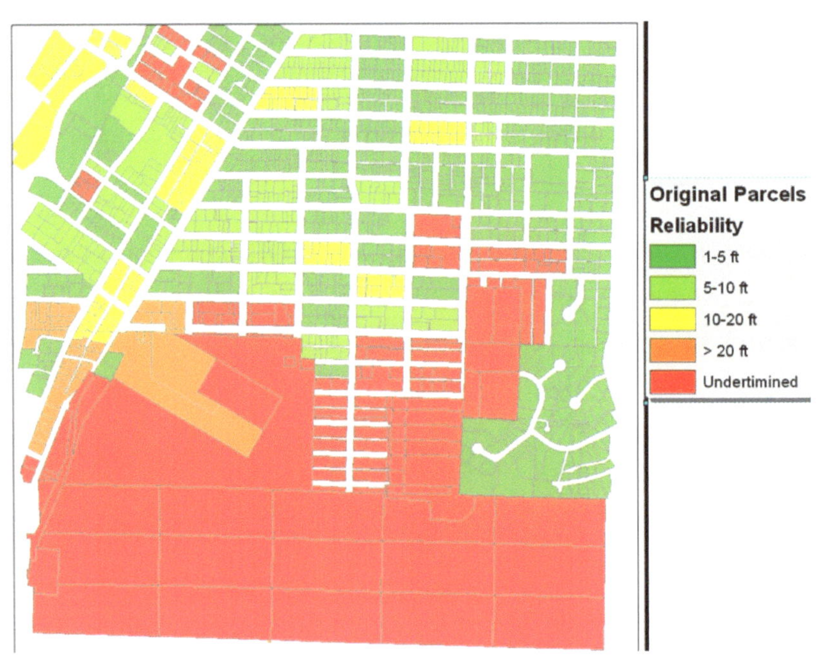

FIGURE 103 RELIABILITY OF ORIGINAL PARCELS

The results of the assessment were input to a GIS in order map the magnitude and distribution of the spatial accuracies across the project area. Error estimate values were entered into the parcel database for each parcel in the pilot area. The parcel errors were aggregated (averaged) to blocks as well because all adjustments were to be performed on a block by block basis. The adjustments were performed by rotating and translating blocks of parcels into proper alignment without changing the mapped geometry of parcels within the blocks.

The origin of the errors could be any number of things ranging from not having sufficient data during the conversion process, drafting errors in the original assessor maps, not having enough survey control, digitizing errors during the conversion, and others. In some cases (and right-of-way issues is one of the bigger examples of this) there are a lot of conflicting bits of information that must be sorted out by a professional land surveyor. Where there are compelling business reasons to do so, time and energy may be devoted to resolving some of the harder issues. However, for the majority of the issues described, a combination of a bit of data entry, field work and cartographic manipulation can improve the alignment issues significantly. The decision on how to proceed and how much time and money to spend on research and adjustments, will depend on how accurate the data need to be for the intended use(s) of the data. For the purposes of this study, a combination of methods for improving the data were tested and evaluated for effectiveness and cost, as described in the consulting surveyors' reported are excerpted below.

PHASE 2 – ACCURACY IMPROVEMENT

METHOD A – Fit parcels to large-scale orthophotography

Random field surveys were conducted using GPS in survey grade RTK operation mode. Survey ties were made to the following items:

- The county's existing orthophotography photo control network points.
- Points identifiable from the county orthophotography, such as fence corners, sidewalk corners, street intersections.
- Property corner pins on block corners.
- Curb lines for controlling blocks.

RESULTS

Original orthophoto control ties checked within 2-3 cm of the published coordinate values. Survey measurements of photo identifiable points matched the calculated orthophoto coordinates (same ID point) at less than 1 foot (30 cm) difference between the two. This provided assurance the orthophotography was fitting with the field surveys. Property pins were recovered and tied to assist in best fitting the parcel mapping. Curb line ties were used in establishing street centerlines and right-of-way lines to assist in best fitting the parcel mapping.

METHOD B – Fit Parcels within road rights-of-way define blocks

Using field survey data and the record plat documents, road and street centerlines and right-of-way lines were defined for those streets where data were available although not all of the right-of-way documents were recovered or available. These roadways defined blocks within which parcels were constrained to fit.

RESULTS

Many improvements to the data were made using this method. The results are shown on the chart in Figure 104 Comparison of parcels accuracy before and after adjustment.

TIME & EFFORT STUDY -

CAD computations only - estimated @ 30-45 minutes/block

Field surveys and CAD computations - estimated @ 60-90 minutes/block

METHOD C – Coordinate geometry fit

Coordinate geometry was calculated from parcel creation documents (deeds and plats) around one subdivision and controlled by property survey ties to property corners.

RESULTS

Significant shifts of 20 feet to 40 feet were identified (6m to 12 m) for the street right-of-way lines and property lines around the cul-de-sac area based on record plat measurements from survey tied property pins versus the original parcel mapping.

Conclusion

Overall, as seen in the chart in Figure 104 Comparison of parcels accuracy before and after adjustment, the parcel accuracies were improved on most data, with the majority of the parcels now falling in the 1 foot-5 foot (30 cm – 1.5 m) range. The resulting accuracies of the

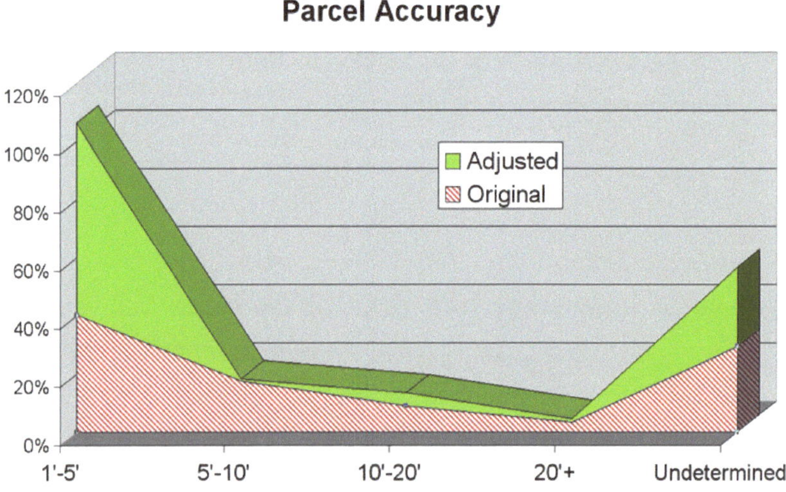

FIGURE 104 COMPARISON OF PARCELS ACCURACY BEFORE AND AFTER ADJUSTMENT

adjusted parcels were mapped for communicating the distribution and location of the results (See Figure 105 Accuracy of adjusted parcels).

Considering that the original conversion relied heavily on heads-up digitizing rather than COGO in most of the original pilot area, and that only a sporadic amount of COGO was done to correct obvious errors the line work, a very large percent of the parcels fell within the 1-5 foot range of after corrections were performed. That is not to say that the line work meets an absolute accuracy in that range, but that when tested against the photography, the parcel lines align very well and will meet the needs of most all, but the most exacting (engineering type) applications.

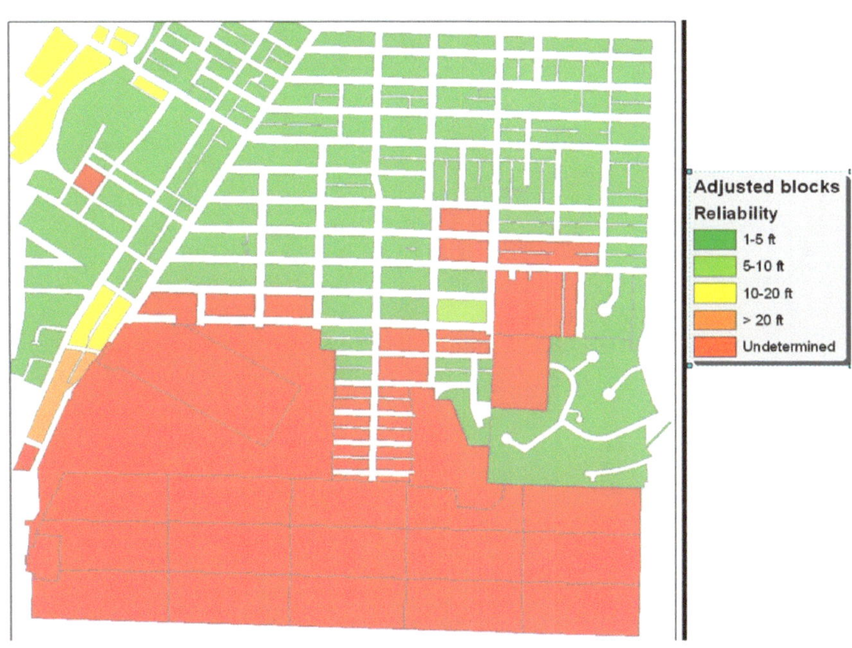

FIGURE 105 ACCURACY OF ADJUSTED PARCELS

In addition to accuracy assessment, and accuracy improvement study, this project helped to quantify the costs of doing the two phases (assessment and improvement) of work. The time analysis results are listed above. For this study, in this area, the cost averaged around $5 per parcel to perform the rigorous assessment and around $10 per parcel to improve the accuracy. The study also demonstrated that the cost to improve the accuracy from greater than 20 feet (6 m) to 5-10 (1.5 m – 3 m) feet was about the same as improving it to 1-5 feet (30 cm – 1.5 m). It is important to note that the amount of work required will vary greatly from area to area and depends on more variables than can be enumerated here. The quality of the original surveys and legal descriptions, the number of conflicting claims, the size and density of parcels, the level of clarity in chain of title, whether or not there are mineral surveys (where are notorious for conflicting and vague descriptions), varying right-of-way widths, and other things all play a role in the how well things will fit together in any given area. The issues that crop up in the source data usually are not resolved in parcel mapping, as that is rarely the intent, although those issues all contribute to the ease or difficulty in improving the spatial accuracy of parcel data.

CHAPTER 33 SURVEYING TO IMPROVE GIS SPATIAL ACCURACY

Surveyors are working to improve the spatial accuracy of GIS layers in Montana.

Because Montana based most of its cadaster (as shown in Figure 106 Montana cadastral website) upon the Bureau of Land Management's Geographic Coordinate Database (GCDB), the process for improving the spatial accuracy of the GIS cadaster first requires upgrades to the accuracy of the GCDB. Montana's surveyors play a significant role in the GCDB enhancement projects where the ultimate goal is to improve the spatial alignment of GIS layers.

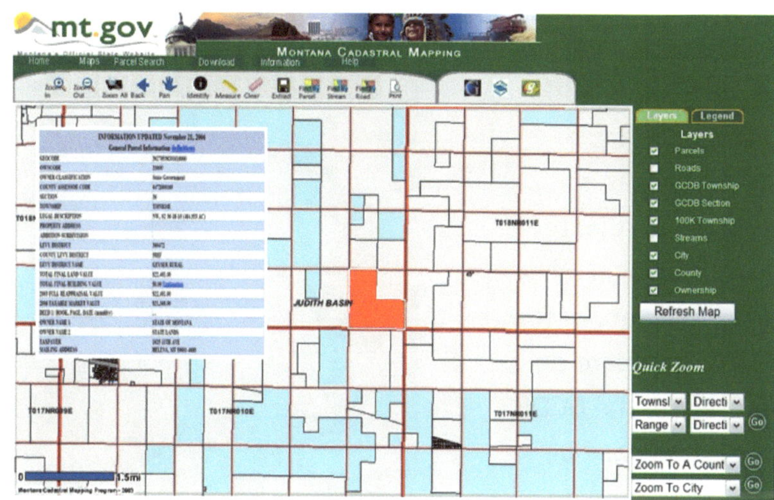

FIGURE 106 MONTANA CADASTRAL WEBSITE

BACKGROUND:

Although Montana is the fourth largest state in land area at more than 145,552 square miles, and has a population density of about 6.2 people per square mile, it is one of the few states in the nation to have a statewide cadastral GIS. The number of parcels in the Montana Cadastral dataset is about the same as the number of people in the state – a little more than 900,000. The annual economic benefits of the Montana Cadastral dataset average in the million dollar range - an obvious demonstration of the value of this system. The benefits accrue from the variety of applications that one can do against a statewide cadastral dataset and the savings in time and money realized by the availability of the data via the World Wide Web (www.cadastral.mt.gov) [see Figure 106 Montana cadastral website]. Nevertheless, the methodologies used to create this statewide cadaster in a few short years in the 1990's resulted in less than optimum spatial accuracy in some areas of the state. "Less than optimum" means that certain applications fail in some areas, such as GPS derived address points not falling in the correct parcel or calculations for impermeable area are difficult to apply uniformly due to the spatial misalignment between GIS layers.

Fortunately, spatial accuracy of GIS layers can be improved and surveyors are part of that process. Montana has a spatial accuracy improvement program dedicated to enhancing the spatial accuracy of the cadastral layer in problem areas.

SOURCES OF ERRORS

Errors arise from a variety of causes. Because Montana based its cadastral framework layer on the Bureau of Land Management's Geographic Coordinate Database (GCDB) in most areas, the accuracy of the parcels is dependent on the accuracy of the GCDB, thus the parcels can be no more accurate than the GCDB. The BLM constructed the GCDB by digitizing coordinates off the US Geological Survey's Topographical maps, supplemented in some areas, by GPS derived coordinates and/or data entry of recent recorded survey bearings and distances. The majority of coordinate pairs of the GCDB are approximations at best. Additionally, other sources of error lie in the reliability of the source data (assessor maps that were hand drafted over the course of decades), and the processes used to digitize the parcels. The digitizing process may have inadvertently introduced errors.

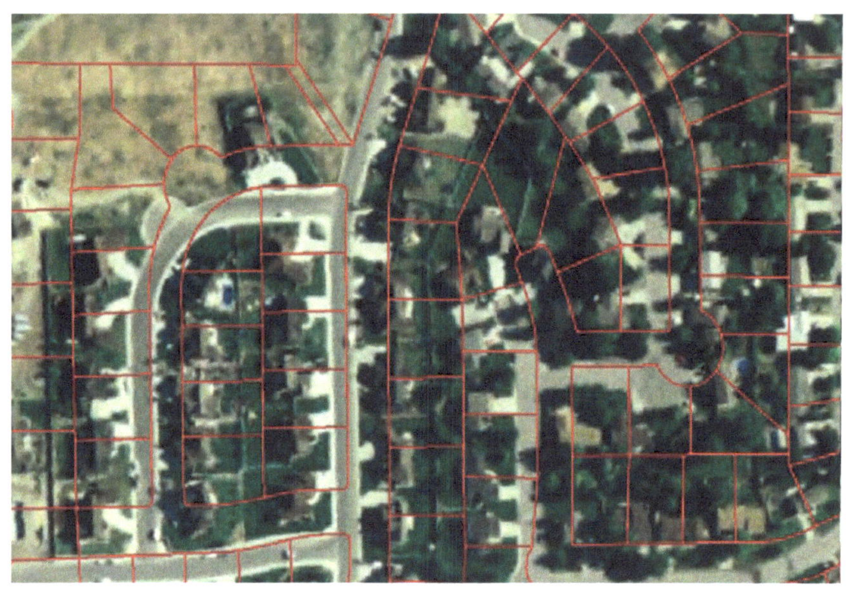

FIGURE 107 Example misalignment of GIS layer against photography

In practice, best results come from minimizing the error in areas where the density of parcels (and other GIS layers) is highest, that is, the higher the parcels density is, the smaller the spatial error should be. Nevertheless, in Montana, the spatial error varies somewhat randomly so we have some urban areas (high density) with very large errors [see Figure 107 Example misalignment of GIS layer against photography], and some rural areas (such as where one whole section might be a single parcel - more or less one square mile) with relatively high accuracy. [See Figure 108 Accuracy of the control layer compared to density of GIS layers)

Regardless of the source of error, one can improve the spatial accuracy in a couple of ways. For rural areas, which are the majority of the state, the process to improve the parcels starts by improving the GCDB. To improve the GCDB, one must develop coordinates that are more accurate - usually by field observations (such as GPS surveys) or by entering, recent, (and presumably more accurate) survey measurements. After entering new coordinates into the GCDB, the GCDB is readjusted to fit the new control, and then the parcels are readjusted to fit the new GCDB. Surveyors and survey measurement data are the key to this process. Surveyors perform the field survey to generate new coordinates, or may even already have coordinates on some Public Lands Survey System corners (tied to the National Spatial Reference System). Surveyors also perform the GCDB re-adjustments, under the guidance of the BLM.

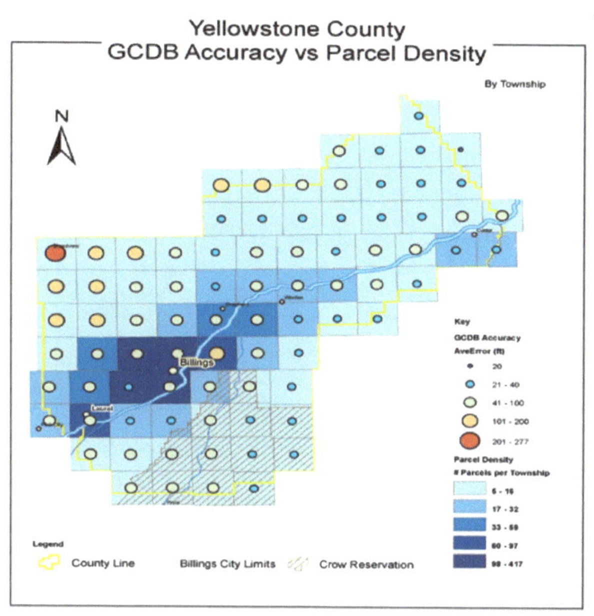

FIGURE 108 ACCURACY OF THE CONTROL LAYER COMPARED TO DENSITY OF GIS LAYERS

Montana received a federal appropriation through the US Department of the Interior; Bureau of Land Management to maintain and enhance the accuracy of GCDB based cadastral framework data. The federal appropriation enabled Montana to enhance the spatial accuracy of those areas of the state where there are gross errors

THE PROCESS

RESEARCH AND PLANNING

The first step is to research corner records, certificates of survey & plats, and BLM records to identify PLSS corners that are suitably monumented and that are likely candidates for coordinating with GPS. Common practice is to completely surround a project area with survey control in order to constrain the adjustment within the project area, and to be able to control the adjacent townships. All corner records retrieved are scanned into TIFF format, given the appropriate GCDB ID, and the relevant information added to a geodatabase and referenced to the corner location so that searches are readily performed, and maps easily produced showing the location of monumented corners that have corner records associated with them. In order to discover whether or not other coordinate data are available on the PLSS stakeholder meetings are held with local engineers and surveyors in private and public organizations (such as engineering firms, surveying firms, utility companies, the Bureau of Land Management, the US Forest Services, the Bureau of Reclamation, etc.). If they have existing survey control on the PLSS that is tied to the National Spatial Reference Systems (NSRS) which they are willing to contribute to the project then their coordinate data can also be used to control the GCDB re-adjustment. In some instances, such as when no reliable monumentation exists or access is cost prohibitive, then data entry of existing recorded survey, plat, and deed information may be more a cost-effective means to improve GCDB coordinate values.

As a planning tool, a map is produced showing where possible PLSS corner candidates are, based on the GCDB reliability estimates (need), the location of existing control coordinate data, the suitability for field ties of PLSS corners (existing corner record indicating a suitable monument in locations that will strengthen the GCDB township, and minimal access and safety issues). Based on that information a survey plan is developed. [See Figure 109 Survey plan for new control for GIS adjustment]

SURVEY STEPS

The next steps are to perform the survey fieldwork (new corner ties, verify other's coordinates, if any).then reduce, calculate and analyze survey data. Field surveys are performed primarily using Global Positioning Satellite survey methods to obtain five (5) centimeter or better accuracy (typical results) on control points. All coordinate information obtained is referenced to the National Spatial Reference System (NSRS) and entered into the GCDB control files as geographic coordinates (NAD 27).

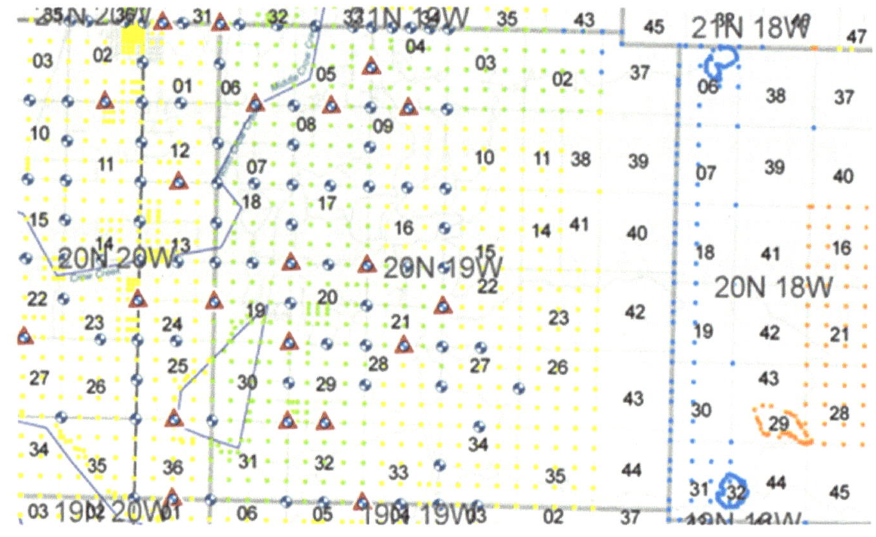

FIGURE 109 SURVEY PLAN FOR NEW CONTROL FOR GIS ADJUSTMENT

All new and existing control coordinate data are entered into a GCDB control file and each GCDB township is re-adjusted using the BLM's GMM software and approved BLM methods and procedures. The adjustment is analyzed for errors and blunders and re-adjusted if necessary, then submitted to the BLM for review. The BLM then incorporates the re-adjusted GCDB into the agency's framework database.

It is important to note that GCDB error estimates are somewhat subjective, so some estimates may be overly optimistic. Because of this it is possible for the error to appear to increase after the adjustment, even though the coordinates may be *more* reliable. In these instances the accuracy is better but merely appears to have worsened because the original estimate overstated the accuracy. However, it may also be true that the accuracy got worse, which does happen in some areas due to bad survey data or blunders in the GCDB. These errors can usually be isolated and measure taken to improve them. Generally, though one will see an improvement in the GCDB accuracy after new data are entered and the GCDB is adjusted (see Figure 110 Changes in error after adjusting to new control), which is, after all, the point.

PARCEL ADJUSTMENT

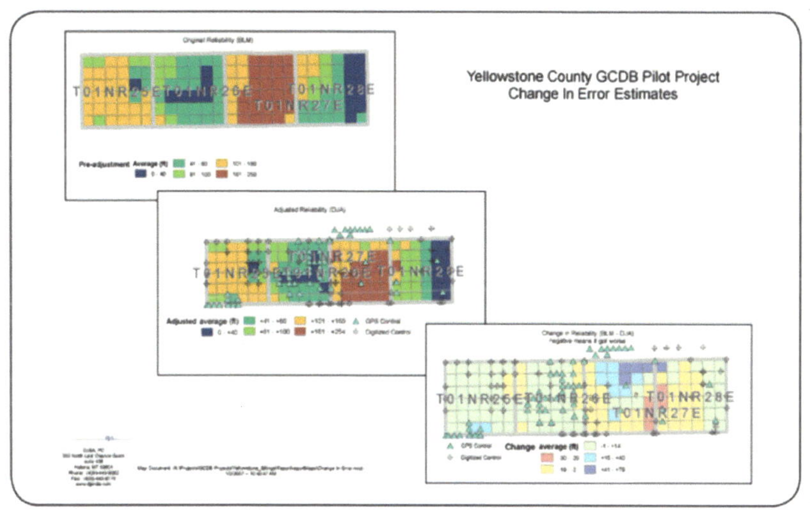

FIGURE 110 CHANGES IN ERROR AFTER ADJUSTING TO NEW CONTROL

After the GCDB is readjusted to the new coordinates, the parcel layer is adjusted by the Montana Department of Administration GIS Bureau, using a least squares adjustment. The least squares adjustment program, developed by Tim Hodson of ESRI, uses the changes in positions of the new GCDB points compared to the old GCDB points for the same corner. This is based on the point identifier (GCDB ID) for points that are found within a user specified search radius. The application automatically finds the matches for each GCDB point (old to new). However, in areas where the new coordinates change significantly (greater than the search tolerance), which sometimes happens, linkages can be made manually. The program then calculates the changes in position and applies those changes to the parcel layer in what is essentially a rubber-sheeting process. Other GIS layers that are linked to the GCDB, such as fire districts, or resource management area boundaries, may also be adjusted in this way.

CONCLUSION

Construction of the Bureau of Land Management's Geographic Coordinate Database was a huge undertaking which is still on-going in some parts of the country and completed in other parts. For those areas that are complete maintenance in the form of accuracy enhancements can occur to help any number of GIS layers fit together better. Surveyors are key players in this process because of the surveyor's skills and knowledge for research, field reconnaissance, field observations, data reduction and analysis. Through the surveyor's effort, the GIS data used by many is improved for all.

PART 7 MISCELLANEOUS ISSUES

In addition to technical considerations, there are social considerations regarding how the activities of GIS and GPS relate to the professional of land surveying. The emergence of sophisticated mapping technologies challenges the level of expertise that traditionally was constrained to the surveying profession. This has created tensions and a bit of anxiety in the surveying profession. In this next part we discuss certain important issues that are highlighted by the collision of traditional surveying activities and 21st century technologies.

CHAPTER 34 WILL LAND SURVEYING BECOME OBSOLETE

Looking back over my nearly twenty-five years of surveying I recall the variety of work that I have had the good fortune to do as a surveyor. At this point in my career, I cannot help but wonder how future changes in technology will affect the work of surveyors. I began my career doing cadastral, construction and forest road surveys. On day one of my surveying career, I helped to set section corners in the temperate rain forest of the Oregon Coast Range in 1977. We used a 300-foot steel tape, and Kern optical theodolite to pull the set from a hub and tack set on a random traverse. After we dug the hole and set the 3 ½ inch galvanized iron pipe with a brass cap on top for the corner, scribed the bearing trees, and wrote up the notes in a field book, we then continued on down the section line blazing trees with a hand axe and setting iron fence posts on-line every 300 feet. We used a hand compass and occasional pre-computed pulls with a cloth tape from previously traversed points to mark the line with blazes and fence posts. It was long hard work in those days, often taking several weeks to brush and traverse a random line, writing all the observations in a field books, transcribing those field notes and entering them into an HP 3800, computing the corner locations for setting the corners, and offsets for setting the posts on-line. All too often the traverse line would have to be brushed again by the time we were ready to go back out to set the corners because it had taken so long to do the transcriptions and calculations. The country was steep and brushy and some of our back sights were fifty or sixty foot stubs. Other times we encountered steep ravines with brush so tall that we couldn't reach to cut it. In those instances we had to break chain to pull the tape down one side of the ravine and up the other instead of across despite the short distance.

FIGURE 111 WRITING SURVEY NOTES

Those days we dreamed of having a way to survey those miles of line through the mountains without having to do all that brushing and lugging gear up and down the mountains for weeks and weeks.

Now here we are in another century and how the technology has changed! Lidar, GPS, electronic data collectors, GIS, CAD, digital photography, online topographic maps, laser measuring devices that fit in your pocket, four wheel ATVs and mobile phones. These technologies have not only changed how we do our work, but they also have changed who can do our work. It really is much simpler now to map just about anything because the tools that we make the measurements with are easier to use. Is this a good thing for the surveying profession or is this a bad thing? Is surveying as a profession becoming obsolete due to changes in technology? Are surveyors losing work due to changes in technology?

There is a lot more mapping of all kinds of things now performed by surveyors and non-surveyors. Additionally, a lot of the mapping that is being done is very detailed and done at a very large mapping scale with excellent precision because of the technology. Back in the last century if a city wanted its manholes mapped, it *had* to hire a surveyor to do it because no one else had the skills and knowledge, or the equipment. But with the advent of GPS a lot of non-surveyors can perform this work and can deliver a satisfactory product in a very short time. The plans for additional civilian GPS signals and changes in GPS and other technologies will eventually result in vastly improved accuracy. Within a few short years we will see real-time centimeter level accuracy. That accuracy will be right there in the palm of the hand of the person holding the unit, regardless of the person's skills, knowledge, background or understanding. What impact will that have on us as surveyors?

It is certainly possible today for a citizen to download county data such as a georeferenced scanned plat of their property, load that into a palm-top computer with integrated GPS, and navigate to their property corner. Now, what they actually find at the "corner" may just be a goat stake or power pole anchor, but people have been mistaking those objects for property pins long before the advent of GPS and GIS. If a property owner wants to build his fence off a goat stake, it's his money and he can spend it any way he wants. Nevertheless, the reality is, as technology advances, his chances of actually finding the real property pin increase. However, I think that we are going to see that more people will rely on technology to give them the answers they seek, instead of calling a professional.

There are certainly numerous worst-case scenarios that we can foresee, but I do not think that the changes are necessarily all bad. One of the major motivations of my work in GIS is my ability to empower people via technology. In local government the focus is primarily for building and maintaining a GIS is so that city and county staff can perform their day to day work more effectively and efficiently. However GIS can directly benefit the citizen, taxpayers, public at large, and businesses.

In recognition of this imperative, public organizations use a variety of avenues to share the GIS data and applications with the public. Some of the ways include publishing paper maps, providing digital copies of the data off the county website, burn DVDs with data along with free GIS software to browse those data and make maps, and with interactive web mapping services that allow the public to make their own maps.

Realtors, engineers, consultants, architects, environmentalisst, and many other professions benefit from the access to and use of these data as well. Because of these changes in technology and the access to so much free and inexpensive data, more people and businesses are making their own maps and mapping their own stuff.

So, it is true, that more and more people are *not* calling a surveyor every time they want something mapped but that is not necessarily a strike against the surveying profession. In my opinion, it actually is beneficial to our profession for the following reasons. First, because of all the issues related to mapping, including issues like projections and coordinate systems and scales, etc. there are many occasions where map data do not fit. Most of the citizens who have stopped in my office with questions about data discrepancies, GPS and other mapping issues, leave with renewed respect for all that stuff surveyors know and can do, and they gain a better understanding of when they do need to call a surveyor. Another reason that empowering the public is beneficial for our profession is that some of the mapping that they do for themselves allows us to focus on the more important stuff. For instance, if a farmer wants to map where his crops are so that he can maximum his application of water, let him do that himself. If that same farmer wants to provide a conservation easement along the west line of his property, he will hire a surveyor to define the location of his boundary and the easement. The farmer knows that mapping his crops is a relatively simple thing that even his hire hands could do, but that boundary surveying requires an expert professional.

So, although we surveyors may see some lost opportunities due to technology changes, overall, I think our profession can become more tightly focused on our greater strengths and ultimately gain more respect professionally. We will all be working smarter rather than working harder.

CHAPTER 35 GIS CERTIFICATION

From time to time, I have heard surveyors say, "GIS is no good, because it was not done by a surveyor." Or that "GIS stands for Get It Surveyed!" Statements like these most often come from surveyors who have little, if any, actual experience with GIS, other than from the outside looking in. Therefore, it comes across as sour grapes. However, such statements beg the question -what is it about GIS that these surveyors are unhappy about? Admittedly, some surveyors perceive GIS as competition that is taking work away from surveyors. This is rarely the case though, since most of GIS has very little to do with surveying. Nevertheless, others may have some legitimate concerns about the quality and accuracy of GIS data, as well as the professional qualifications of the people doing the GIS work. Because GIS is used for an incredible variety of purposes, it is done by people from many different professions. Biologists, planners, wildlife experts, health professionals, law enforcement officers, firemen, and many others get into GIS in order to help them and their colleagues do their jobs better. GIS provides a means for managing their information more effectively and efficiently. Many of these folks get into GIS without much training in mapping, error detection, coordinate systems, projections, deed reading, and other things pertinent to this type of work. They do understand their own data and how to use it, but they may not all know how to properly map the data, or tell when they are not getting accurate coordinates. This is not the case for all folks doing GIS, but is the case for some.

Surveyors do have an important role in developing a few of the layers used in GIS, such as plats and surveys and control points. Surveyors also have an interest in using those same layers and many others, such as soils, wildlife habitat, land ownership, transportation and more. It is that dependence upon the work of others that evokes concerns about the quality of GIS. Anyone who uses GIS wants to trust that the data are complete, accurate, and spatially reliable. Reviewing the Metadata that comes with a dataset is, the best way to determine where the data originated, the appropriate scales for use, how and when the data were converted, and how spatially accurate they are. Metadata needs to be there, and it needs to be reliable itself. This is not always the case. Unfortunately, some GIS professionals do not create Metadata, or create incomplete Metadata, or, in some cases, may even misrepresent the accuracy of the dataset. Lack of Metadata and lack of faith in the Metadata diminishes the reliability of the GIS data, which in turn, diminishes the trust that one has in the person or agency that created or used the data.

There are some surveyors and GIS folks, who contend that a certification program for GIS would elevate the credibility of GIS professionals and increase confidence in GIS data. The arguments are, that a certification process for GIS professionals would introduce a certain, and perhaps measurable, level of professionalism to the field, along with assurances of an ethical code of conduct.

WHAT IS GIS CERTIFICATION?

The Urban and Regional Information Systems Association assigned a committee that dedicated many months to develop a recommendation on certification. URISA is the founding member of the GIS Certification Institute, the organization that administers professional certification for the field and is dedicated to advancing the industry. At the time, URISA's committee members "...felt that a "typical" GIS Certified Professional has the following characteristics:

- a formal degree with a number of specific GIS and GIS-related courses or the equivalent coursework in professional development courses and other educational opportunities;
- At least four years of experience in a position that involves data compilation, teaching, etc. (fewer years if in GIS analysis, design, or programming; and more years if in a GIS user position); and
- A modest record of participating in GIS conferences, publications, or GIS-related events (such as GIS-Day). "

The proposal recommends a point system for various areas of professional experience, contributions to the profession (presenting seminars, writing articles, etc.), and education that is GIS specific.

The interesting thing about their approach is that "Certification is career recognition [author's emphasis] through the evaluation and approval of individuals engaged in a specific occupation or profession. "What this means is that certification would be not be based on competence, but on the presumption of competence that should naturally arise from experience, education and professional contributions. In other words, one would not have to demonstrate competence to be certified. One would only have to demonstrate that one has done a lot of stuff – good, bad, or otherwise.

HOW IS CERTIFICATION DIFFERENT FROM LICENSING?

The URISA proposal intended to establish ethical and professional standards. Professional licensing, on the other hand is oriented toward protection of the public. To become licensed as a surveyor, one must document sufficient experience and education just to be allowed to sit for the examinations, which test one's knowledge and understanding. That is, not only does one need experience and education, but one also must prove a high level of knowledge in order to become licensed. Other factors come into play as well, such as the type of experiences and level of responsibility and ethics, etc. However, the crux of the matter is that one must have experience, education and a certain level of competence and knowledge in order to obtain a surveying license. Testing was considered by the GIS certification committee, but was rejected because GIS embraces such broad technical areas.

WHAT IS THE TIME FRAME FOR GETTING GIS CERTIFICATION IN PLACE?

URISA adopted this proposal during the summer of 2002 and ran a pilot program for a few months to test it out. In order to do that an organization must be created to do the certification. The proposal is on the URISA web site and is open for review and comment at the time of this writing.

ARGUMENTS AGAINST GIS CERTIFICATION

Many of the arguments against GIS certification have been that GIS embraces so many different professions that it is difficult to get anybody to agree what GIS is and what it is not. I can certainly understand these arguments. In my work, I provide GIS support for a city and a county. Our goal is to put GIS, that is, the tools and the data, into the hands of the professionals, so that they can do their work, in a GIS way. That is, the planners, engineers, road maintenance workers, the sign guy, the water quality department, the septic inspectors, firemen, law enforcement officers, and many others, will (and most do) use GIS to do their work. They are not going to do only GIS work. They are going to do the work that they have always done, but they will now use this other tool (GIS) to help them do it. Because of this diversity in backgrounds and use of GIS, it is difficult to agree on which skills, knowledge and experience is necessary to be a competent GIS professional, or even what a competent GIS professional is.

HOW DOES CERTIFYING GIS EXPERTS AFFECT SURVEYING?

A certification program for GIS professionals that ensured certain levels of competence and professional ethics, I think addresses some of the concerns that surveyors have about GIS data and the way things are done in GIS. However, I doubt that surveyors would find those assurances in the URISA proposal. It would probably be quite difficult to come to a reasonable agreement, even within the surveying community, on what GIS is or is not; how GIS ought to be done; and who ought to do it, let alone agreeing on the kinds and levels of skills, experience and knowledge required for certification. Nevertheless, in the absence of anything else, we have no guidance to help us ascertain how good a GIS professional is at what he or she does.

I have heard the suggestion that instead certifying people, we should certify the data. There is a lot of sense in that argument, because most of the issues and concerns about GIS are about the data. However, if we certify data, who would do the data certification, and who would certify that the data certifiers were competent to certify those data?

APPENDIX

REFERENCES

Chrisman, N. R. (1997). Exploring geographic information systems. New York: J. Wiley & Sons.

FGDC Website http://www.fgdc.gov/nsdi/nsdi.html (February 2013)

National Atlas website: http://www.nationalatlas.gov/articles/mapping/a_projections.html (February 2013)

Idaho State University Geology website: http://geology.isu.edu/geostac/Field_Exercise/topomaps/distortion.htm (February 2013)

UNAVCO website: http://facility.unavco.org/project_support/tls/tls.html (February 2013)

USA General Services Administration document – from www.gsa.gov (accessed February 2013): http://www.gsa.gov/graphics/pbs/GSA_BIM_Guide_Series_03.pdf

GLOASSARY

ALIQUOT PARTS	The standard subdivisions of a PLSS section by quantities, e.g. 160 acres or 40 acres, etc. that results from splitting the rectangular geometry in halves, quarters, etc.
ATTRIBUTES	The text and numeric database values associated with GIS feature geometry.
BLM	Bureau of Land Management of the United States of America, Department of the Interior.
CAD	Computer Aided Drafting (or Design) is the computer software technology for creating digital objects.
CADASTRAL	Of or pertaining to cadaster which is a system of records and descriptions of land ownership and interests.
CLOSING CORNER	is a corner established where a survey line intersects a previously fixed boundary at a point between corners.
COGO	Coordinate Geometry is a geometric means of creating GIS features, such as drawing lines using coordinates of end points, or by calculating coordinates of endpoints from the measure and angle of a line (e.g. bearing and distance).
CONTROL POINT	A fixed location with coordinates that are known with high reliability.
GCDB	The Geographic Coordinate Database is the Bureau of Land Management's computer representation of the Public Lands Survey System. The GCDB is based on coordinates for each PLSS corner. PLSS corners are angle points, points on line, meandered points, reference points and other points that define the shape and location of the PLSS rectangular system, and may also include mineral surveys, homestead entries, meanders along navigable waterways, and other survey information.
GIS	GEOGRAPHIC INFORMATION SYSTEM- Computer hardware, software, people, processes, and data that comprise a graphical interface to information that has a location component.
GPS	Global Positioning Satellite System is satellite and ground based system of hardware and software used to locate positions on the earth via GPS receivers. Positions are calculated based on the length of time that the GPS signals take to go from the satellites to the GPS receiver.
MEANDERS	The approximate location of water bodies (such as lakes, rivers, swamps, etc.) as mapped during the original Public Lands Surveys. The meanders do not necessarily depict water body edges as they are today, nor do the meanders have any legal bearing.
MISCLOSURE	The failure of survey geometry to form a completely closed polygon due to the starting and ending points having differing coordinates.
OFFSET CORNERS	Duplicate corners of Public Lands Survey System where the ending points of one township are located in a different location than the corners of the adjacent township where in most instances the two corners are coincident. Most offset corners occur along the north or south lines of townships, but may be in other places as well/

NON-GCDB	A non-GCDB PLSS layer is one that is not based on the GCDB. This could be a PLSS representation from the US Census Bureau TIGER files or one that was created by an agency that digitized PLSS tic marks on a USGS Topographical Map Series.
NON-PLSS	In the context of this document, this means any boundary that is not directly referenced to the PLSS. Typically this would be metes and bounds descriptions, roadways, rivers, and other features that are boundaries.
PLSS	Public Lands Survey System is the land allotting system design by President Thomas Jefferson for the orderly disbursement of the western public lands. Also known as the rectangular system, the PLSS is comprised of township blocks approximately six miles by six miles square that are further divided into sections of land that are approximately one square mile in area.
REGISTER to the GCDB	As used in this document, a data set that is registered to the GCDB is one for which the GCDB is used as the mapping control, that is, the GIS layer has at least some of its features coincident to the GCDB points, lines, or polygons. For example, if a parcel boundary follows a section line, and the GIS feature of that parcel uses GCDB section corner points or GCDB section lines, or GCDB section polygons was used to define where that section line is located, then that parcel is registered to the GCDB. This could be done by snapping the parcel vertices to the GCDB points, or by using a GCDB line to create part of the parcel boundary.
TIGER	Topologically Integrated Geographic Encoding and Referencing system are GIS products that were produced by the United States Census Bureau as a mapping and analysis aid for collecting and understanding population information within the United States of America/
TOPOLOGY	As used in GIS, is the measure of connectivity among features of a dataset or between features of differing datasets.
WGA	Western Governors Association

INDEX

P

R

S

T

V

W

TABLE OF FIGURES